DEVELOPMENTS IN
POLYMER CHARACTERISATION—4

CONTENTS OF VOLUMES 2 and 3

DEVELOPMENTS IN POLYMER CHARACTERISATION—4

Edited by

J. V. DAWKINS

Chemistry Department, Loughborough University of Technology, UK

APPLIED SCIENCE PUBLISHERS
LONDON and NEW YORK

APPLIED SCIENCE PUBLISHERS LTD
Ripple Road, Barking, Essex, England

Sole Distributor in the USA and Canada
ELSEVIER SCIENCE PUBLISHING CO., INC.
52 Vanderbilt Avenue, New York, NY 10017, USA

British Library Cataloguing in Publication Data

Developments in polymer characterisation.—4.—
(The Developments series)
1. Polymers and polymerization—Analysis—
Periodicals
I. Title   II. Series
547.7'046'05   QD139.P6

ISBN-13:978-94-009-6630-7 e-ISBN-13:978-94-009-6628-4
DOI: 10.1007/978-94-009-6628-4

WITH 119 ILLUSTRATIONS AND 9 TABLES

© APPLIED SCIENCE PUBLISHERS LTD 1983
Softcover reprint of the hardcover 1st edition 1983

Typeset in Great Britain by Keyset Composition, Colchester

PREFACE

This volume includes reviews on tackling polymer characterisation problems and on developing specific characterisation techniques. The first two chapters and the last chapter describe progress in providing characterisation information for polymers containing long-chain branching, for polymer blends, and for polymers having preferred orientation. The remaining chapters review progress in individual techniques, showing with examples the characterisation results which may be obtained.

It is recognised that the degree of chain branching which can evolve in some polymerisation processes can have a marked effect on the flow properties of a polymer, and therefore on polymer processing behaviour. In the first chapter the characterisation of long-chain branching from measurements of the molecular size and molar mass of a polymer in dilute solution is outlined. It is indicated that a complete characterisation of branching requires the combined use of several techniques, emphasising in particular recent developments involving gel permeation chromatography.

Thermal analysis and infrared spectroscopy are widely used techniques in polymer characterisation. Both techniques can provide, very quickly, significant results with readily available instrumentation. This is illustrated by the review of the characterisation of polymer blends by thermal analysis in Chapter 2. An assessment of blend morphology, which influences the behaviour of a material consisting of two or more polymers, is presented in terms of transition temperatures. Conventional infrared spectroscopy involves dispersive spectrometers which do not always provide accurate information on composition and structure for complex polymeric materials. The application of interferometric techniques to infrared instrumentation has increased sensitivity so that many characterisation problems which were formerly difficult may now be accomplished. Furthermore, analysis of the spectrum with computerised data-processing routines permits the resolution of overlapping peaks and the identification of weak bands. These developments in Fourier transform infrared spectroscopy are given in Chapter 3.

Studies of molecular relaxations in polymeric systems have attracted considerable interest because information on time-dependent behaviour is required when identifying the commercial application of a polymeric material. Traditionally, relaxation phenomena in polymers have been examined by means of dielectric and dynamic mechanical techniques. However, in a comprehensive investigation of a material, additional techniques are necessary in order to provide information, for example, on relaxations over a wide frequency range and on the dynamics of specific groups in a polymer chain. Progress in studies of relaxations and dynamics by neutron scattering and ultrasonic propagation are included in Chapters 4 and 5.

Polymer properties may be significantly influenced by the degree of preferred orientation introduced by drawing or other forming processes. In a semi-crystalline polymer characterisation of both the crystallite orientation and the orientation of chain segments in the amorphous regions will be required. Several experimental methods are available for evaluating orientation but differ in the type of information they provide. In Chapter 6, a general approach in terms of an orientation distribution function is presented for the characterisation of preferred orientation. This approach demonstrates how the complete orientation distribution function may be obtained for the crystallites from experimental data determined by X-ray diffraction and how other experimental methods only provide information on some of the moments of the distribution function.

J. V. DAWKINS

CONTENTS

LIST OF CONTRIBUTORS

J. R. FRIED
College of Engineering, Department of Chemical and Nuclear Engineering, University of Cincinnati, Cincinnati, Ohio 45221, USA.

J. S. HIGGINS
Department of Chemical Engineering and Chemical Technology, Imperial College, Prince Consort Road, London SW7 2BY, UK.

B. JASSE
Laboratoire de Physicochimie Structurale et Macromoleculaire, Ecole Superieure de Physique et de Chimie de Paris, 10 rue Vauquelin, 75231 Paris, France.

HIROMICHI KAWAI
Department of Polymer Chemistry, Kyoto University, Kyoto 606, Japan.

SHUNJI NOMURA
Department of Textile Engineering, Faculty of Textile Fibres, Kyoto University of Industrial Arts and Fibre Technology, Kyoto 606, Japan.

RICHARD A. PETHRICK
Department of Pure and Applied Chemistry, University of Strathclyde, Thomas Graham Building, 295 Cathedral Street, Glasgow G1 1XL, UK.

TH. G. SCHOLTE
*Central Laboratory, DSM, Research and Patents, PO Box 18, 6160
MD Geleen, The Netherlands.*

Chapter 1

CHARACTERISATION OF LONG-CHAIN
BRANCHING IN POLYMERS

Th. G. Scholte

Central Laboratory, DSM, Geleen, The Netherlands

SUMMARY

The determination of long-chain branching (LCB) of polymers is in most cases performed by evaluation of both molecular dimensions and molar mass. Several experimental techniques, e.g. light scattering, diffusion, sedimentation, viscometry and gel permeation chromatography, can furnish the necessary data. Indices of LCB in use are the ratios for branched and linear macromolecules of the mean square radii of gyration (g), of the hydrodynamic radii (h) and of the intrinsic viscosities (g'). A survey is given of how these indices can be determined from the experiments, how they are interrelated, and how they are connected with the LCB structure and the number of long side chains. The complications occurring with samples polydispersed in molar mass and in degree of branching are also considered.

1. INTRODUCTION

The term 'long-chain branching' (LCB) is used to describe the configuration of a polymer molecule possessing side chains with a degree of polymerisation of the same order as that of the main chain. Some polymerisation processes, for instance anionic polymerisation, produce linear chains. Other processes, e.g. radical polymerisation, may yield strongly branched molecules. A well known example of the latter type of material is low density polyethylene (LDPE), in which intermolecular chain transfer causes side chains to build up on a chain formed earlier.

1

Long-chain branching very strongly influences the dimensions of the molecules, and hence not only the viscosity in solution, but also the flow properties and, consequently, the behaviour during processing. At the same time, it influences the material properties. In view of this, it is important that good methods of detection are available to determine both the nature and the degree of long-chain branching in a polymer.

The occurrence of long side chains in macromolecules has long been known,[1] as has the main influence of long-chain branching on various polymer properties.[2] The synthesis of well defined branched polymers and the correlation of their structure with the properties they exhibit has brought an important advance in the study of LCB. However, in many cases a complete characterisation of LCB is still impossible. This is due on the one hand to the impossibility of making an exact determination of the branched-chain structure in a number of cases, and on the other hand to the fact that for various structures the relation between the properties determined on the polymers (e.g. hydrodynamic data) and the degree of LCB is not (yet) fully known. In such cases one has to make do with experimentally defined branching parameters, which bear a not yet fully known relation to the actual chain-branching structure.

In most methods used to determine molar masses, quantities are measured which depend not only on the molar mass but also on the LCB structure. Knowledge of this structure is therefore required to determine molar masses and also molar mass distributions correctly.

We shall not deal here with the important subject of the influence of long-chain branching on the flow behaviour and other material and product properties of polymers. For a treatment of this subject the reader is referred to some recent reviews (e.g. ref. 2).

2. MOLECULAR DIMENSIONS OF LONG-CHAIN BRANCHED POLYMERS

The way in which polymerisation is effected and long side chains are built up determines the branching structure of the macromolecules formed; hence, the structure may differ widely from one product to another. The most direct result of long-chain branching is a drastic reduction in molecular dimensions. To obtain quantitative information on this point we shall first look at the influence of branching on the unperturbed dimensions.

2.1. Unperturbed Dimensions

If the chain structure is known, it is possible to calculate, for the unperturbed state, the average molecular dimensions from segment lengths and the number, sizes and interdependence of subchains, using random-flight statistics. A dimensional parameter that is often used is the so-called mean-square radius of gyration, $\langle S^2 \rangle$.[3] A factor g is used to indicate the ratio of the $\langle S^2 \rangle$ of a branched macromolecule to the $\langle S^2 \rangle$ of a linear one having the same number of chain segments (the same molar mass):

$$g = \langle S^2 \rangle_{br} / \langle S^2 \rangle_{lin} \tag{1}$$

On the basis of equations given by Kramers,[4] by Zimm and Stockmayer,[5] and by Kataoka,[6] various authors[5, 7, 8] derived expressions for the factor g for the three principal types of long-chain branching, viz. the star type (one branch unit, from which more than two branches radiate), the comb type (one main chain with side chains attached to it in several places), and the random type, in which branch units are randomly distributed over all chain segments (a random mixture of all structural isomers with a given number of branch units). The factor g can be calculated for a uniform distribution of subchain lengths (uniform length of all subchains between two successive branch units or between a branch unit and the end of a chain), as well as for a random distribution of subchain lengths. With star-shaped configurations g then depends exclusively on the functionality of the sole branch unit (number of branches). In comb-type and random-type molecules, g depends on the number of branch units and their functionalities, which usually have a value of 3 or 4.

For a star with functionality f and random subchain lengths, g amounts to:

$$g_r \text{ (star)} = \frac{6f}{(f+1)(f+2)} \tag{2}$$

For a comb-shaped molecule with m f-functional branch units and p subchains ($p = (f-1)m+1$) with random lengths, g is:

$$g_r \text{ (comb)} = \frac{6p^2 + (f-1)^2 m(m^2-1)}{p(p+1)(p+2)} \tag{3}$$

For macromolecules having random chain branching with on average m branch units the following equations hold within a 2% limit:[5]

if the branch units are tri-functional:

$$g_r \text{ (random)} = [(1 + m/7)^{1/2} + 4m/9\pi]^{-1/2} \tag{4}$$

and if the branch units are tetra-functional:

$$g_r \text{ (random)} = [(1 + m/6)^{1/2} + 4m/3\pi]^{-1/2} \qquad (5)$$

The branching density or branching frequency, λ, is the number of branch units in a molecule, divided by the molar mass. The number of branch units per molecule, m, therefore, amounts to the product of λ and molar mass M.

In polydisperse samples, the molar mass distribution is also a very important factor. For the case of the most probable distribution of branched polymers containing branch units of any single functionality,[9, 10] Zimm and Stockmayer[5] calculated the average values of g. For random type branching with tri- and tetra-functional branch units the weight average and the z average values of g, expressed in terms of the weight average number of branch units n_w per molecule ($n_w \gg 1$), are:

$$\langle g_3 \rangle_w = \frac{6}{n_w} \left[\frac{1}{2} \left(\frac{2 + n_w}{n_w} \right)^{1/2} \ln \left(\frac{(2 + n_w)^{1/2} + n_w^{1/2}}{(2 + n_w)^{1/2} - n_w^{1/2}} \right) - 1 \right] \qquad (6)$$

$$\langle g_4 \rangle_w = \frac{1}{n_w} \ln (1 + n_w) \qquad (7)$$

$$\langle g_3 \rangle_z = \frac{1}{1 + n_w/3} \qquad (8)$$

$$\langle g_4 \rangle_z = \frac{1}{1 + n_w} \qquad (9)$$

For branched macromolecules with a closed ring structure, g is considerably smaller than for open chains.[3, 5]

It is important to keep in mind that the above expressions hold only for the given distributions of branch units and—where applicable—for the given molar mass distributions. However, for lack of better expressions they are often used for samples about which the only information available is that they are monodisperse or polydisperse and that the molecules show long-chain branching.

In the experimental technique applied in the determination of the molecular dimensions which depend on chain structure, use may be made of dimensional parameters other than the mean-square radius of gyration. The most important among these other parameters is the hydrodynamic radius R (or hydrodynamic diameter or hydrodynamic volume), which has an effect not only on the friction coefficient, but also on the dilute-solution viscosity, and plays a role in gel permeation chromatography (GPC). The LCB index used in this case is the ratio of the hydrodynamic radius of a branched macromolecule to that of a linear one having the same molar mass:

$$h = R_{br}/R_{lin} \qquad (10)$$

For the most common types of chain branching (star, comb and random types) the factor h has been calculated.[8, 11, 12] This parameter will be dealt with in the discussion of the relevant techniques.

2.2. Dimensions in Good Solvents

Most polymer–solvent systems do not produce ideal or pseudo-ideal solutions. As is known, macromolecules dissolved in a good solvent have dimensions larger than the unperturbed dimensions in a theta solution. For the mean-square radius of gyration this may be expressed by means of the expansion factor α, introduced by Flory:[13]

$$\langle S^2 \rangle = \alpha_S^2 \langle S^2 \rangle_0 \qquad (11)$$

Hence, the factor $g = \langle S^2 \rangle_{br}/\langle S^2 \rangle_{lin}$ for real macromolecules is not equal to the ratio between $\langle S^2 \rangle_{br}$ and $\langle S^2 \rangle_{lin}$ for unperturbed molecules, unless the α_S of the branched molecule is exactly equal to the α_S of a linear molecule with the same molar mass. Calculations and measurements of the $\langle S^2 \rangle$ of molecules with star-type and comb-type branching show that α_S here has a greater value than it has for linear macromolecules.[3, 14–16] Hence, the dimensional reduction brought about by long-chain branching is smaller than the corresponding reduction in unperturbed molecules. However, other calculations and measurements of the α_S of macromolecules with star-type branching lead to the opposite conclusion.[17, 18]

3. EXPERIMENTAL METHODS FOR DETERMINING LONG-CHAIN BRANCHING IN MACROMOLECULES

In general, analysis of long-chain branching sets out to clarify the overall structure of a polymer sample, i.e. to obtain information on both the type and the degree of branching (number of branch units or number of side chains per molecule). If no information is available, determining the type of branching is a difficult and often impossible task; it may be very complicated even if the way in which the polymer was prepared is known.[19]

Most of the methods that are used to determine the degree of long-chain branching in polymers in essence comprise the determination of the molecular dimensions in a solution. Various eligible methods yield parameters which depend both on molar mass and on molecular size, so that at the same time they give information on the molar mass.

A direct method for determining the radius of gyration is the measurement of light-scattering. Also the coefficient of friction of a dissolved macromolecule is a good parameter, as it is strongly dependent on molecular dimensions. This friction constant can be determined both by diffusion and by sedimentation in the ultracentrifuge. Further, the intrinsic viscosity of a polymer in solution is a parameter that is directly dependent on molecular dimensions.

One of the principal methods for determining molar masses and molar mass distributions is gel permeation chromatography (or, by a better name, size exclusion chromatography) which in most cases amounts to the determination of molecular size. This means that the method is also very suitable, e.g. in combination with a determination of absolute molar mass, for obtaining information on the reduction in molecular size as a result of branching.

Although the second virial coefficient of a polymer in a good solvent depends on branching,[3, 20–22] it has not often been used as a means to determine LCB, partly because intrinsic viscosities, for instance, can be measured with much greater accuracy. Finally, the rheological behaviour of polymer melts is strongly dependent on long-chain branching, so that the viscoelastic properties of such a melt can also provide information on LCB.

Nuclear magnetic resonance can be used to clarify the microstructure of many types of molecules. High-resolution ^{13}C NMR, developed in recent years, can, for many macromolecules, give information on the number of side chains and—up to a certain length—also on the lengths of such chains.

With polydisperse samples the methods for determining LCB can be applied to the whole sample. In most cases, however, more comprehensive information is obtained if a sample is first fractionated and the fractions then examined. As many fractionating techniques separate according to both molar mass and degree of branching in a not completely known way, fractionation by preparative GPC, where the separation is based on molecular size only, may offer advantages here.[23, 24]

3.1. Determination of Molecular Dimensions by Means of Light Scattering

The light-scattering technique provides the possibility to obtain information on the shape and dimensions of macromolecules from the angular dependence of the intensity of the scattered light. It follows directly from theory that, for concentrations approaching zero and small values of the scattering angle θ, the factor indicating the reduction in scattering intensity caused by interference of the light scattered by the various units in a macromolecule, $P(\theta)$, depends exclusively on the scattering angle and on the radius of gyration:[25]

$$\lim_{\substack{\theta = 0 \\ c = 0}} P(\theta) = 1 - \frac{16\pi^2}{3\lambda^2} \langle S^2 \rangle \sin^2 (\theta/2) \qquad (12)$$

(λ = wavelength of the light in the medium).

For a polydisperse sample $P(\theta)$ is correlated in this way with the mass average of the product of molar mass and $\langle S^2 \rangle$ divided by M_w. This quotient is called the z-average radius of gyration and for Gaussian coils it is, indeed, the mean-square radius of gyration of molecules whose molar mass is equal to the M_z of the whole sample. In this kind of average the heaviest molecules make the greatest contribution to $\langle S^2 \rangle$, so that $P(\theta)$, for branched molecules as well, is very strongly dependent on the dimensions of the heaviest molecules, which in most cases also possess the highest degree of long-chain branching.

Experimentally, the determination of $\langle S^2 \rangle$ by light scattering may meet with several difficulties. Especially with lower values of the radius of gyration, the slope of the corresponding line in the Zimm plot is rather weak, which means that for samples with $M_z < 10^5$, $\langle S^2 \rangle$ can be determined with only low accuracy in the case of linear molecules and with even less accuracy for branched (and hence smaller) molecules. With strongly polydisperse samples the $P(\theta)$ vs $\sin^2(\theta/2)$ function for low values of θ is sometimes strongly curved,[25-27] which also makes a correct determination of $\langle S^2 \rangle$ very difficult. The presence of dust and cross-linked polymer components (e.g. microgel) has an even greater influence in the determination of $P(\theta)$ than in the determination of M_w.

For a determination of the LCB index g, it is necessary that two quantities be known, viz. $\langle S^2 \rangle_z$ and M_z, the experimental determination of either of which is rather inaccurate. M_z can be determined directly only from measurements of the sedimentation equilibrium;[28] an estimate derived from M_w (light scattering) and M_n (osmometry) requires knowledge of the molar mass distribution, and is therefore rather unreliable in many cases. In addition, it is desirable that measurements be performed at the theta temperature. For branched polymers, however, the theta temperature is not readily available. Also, under theta conditions there is a much greater likelihood of aggregate formation, which may strongly interfere with the measurement of the radius of gyration.

In good solvents the measurement of $\langle S^2 \rangle$ is undoubtedly somewhat easier. When the limiting value of $\langle S^2 \rangle$ for concentrations approaching zero is reduced the intermolecular interaction can be eliminated. The intramolecular interaction, manifesting itself in an expansion of the molecules in a good solvent, does not disappear, however, so that for the difference in expansion factor (excluded-volume effect) between branched and linear macromolecules a correction has to be applied if g has to be determined for unperturbed dimensions.

3.2. Diffusion and Sedimentation

For very low concentrations, the diffusion coefficient D_0 occurring in Fick's laws can be written as:

$$D_0 = \frac{kT}{f} \tag{13}$$

in which the constant k is R/N_A (R = universal gas constant, N_A = Avogadro's number), and f is the translational friction coefficient.

With a solution of macromolecules a distinction can be made between the free-draining and the non-free-draining case, with a state of partial free draining in between. In the free-draining case, the coefficient of friction of the whole of the macromolecule is the sum of the friction coefficients of the individual segments. When there is no free draining—the more realistic case—f can be rendered, in analogy with Stokes' law for macroscopic spheres, as:

$$f = 6\pi\eta_0 R \tag{14}$$

where η_0 represents the viscosity of the medium (i.e. the solvent) and R is the so-called hydrodynamic radius of the macromolecule.

For linear chains in the unperturbed state it is calculated that:[3, 29]

$$f = \frac{9\pi^{3/2}}{4} \eta_0 \langle S^2 \rangle_0^{1/2} \tag{15}$$

Hence:

$$R = \tfrac{3}{8}\pi^{1/2}\langle S^2 \rangle_0^{1/2} \tag{16}$$

and, because $\langle S^2 \rangle_0$ is proportional to the molar mass M:

$$D_0 = K_D M^{-1/2} \tag{17}$$

Under non-theta conditions R is proportional to $M^{(a+1)/3}$ (a = exponent of the Mark–Houwink equation), so that in this case for the limit of zero concentration:

$$D_0 = K_D M^{-(a+1)/3} \tag{18}$$

For branched macromolecules the hydrodynamic radius, and hence the friction coefficient f, is smaller by the abovementioned factor h, so that:

$$D_0 = h^{-1} K_D M^{-(a+1)/3} \tag{19}$$

with h defined as in eqn. (10). Determination of the diffusion coefficient of a branched monodisperse sample and of a linear sample with the same molar mass gives the ratio between the diffusion coefficients, and hence the factor h. It is necessary to carry out the measurements under theta conditions or to extrapolate towards zero concentration.

For the most frequent types of chain branching the factor h has been calculated for the molecule in the non-free-draining condition.[8, 11, 12] If now the type of branching is known, the number of branch units can, for the unperturbed condition, be calculated from the reduction of the friction coefficient, and hence from the increase of the diffusion coefficient.

For polydisperse polymers, the weight-average diffusion coefficient gives the weight-average reciprocal friction factor, and hence the weight-average value of the reciprocal hydrodynamic radius.

In addition to the classical methods for measuring diffusion coefficients, a new technique for determining diffusion coefficients of polymer molecules in dilute solutions was recently developed, viz. the inelastic light scattering technique.[30–32] An interesting possibility is the simultaneous determination of $\langle S^2 \rangle$ and R from integrated and dynamic light scattering so as to obtain information on the type of branching from the quotient $\langle S^2 \rangle^{1/2}/R$ (or from $g^{1/2}/h$).[31, 32]

Sedimentation is subject to practically the same influence of branching as diffusion. The sedimentation coefficient S is defined as the velocity of the molecules in a centrifugal field divided by $\omega^2 r$ where ω is the angular velocity and r is the distance to the axis of rotation. This velocity is equal to the centrifugal force exerted on the macromolecule divided by the friction coefficient. From this it follows that:

$$S = \frac{M(1 - v\rho)}{N_A} \frac{1}{f} \tag{20}$$

with $(1 - v\rho)$ being the buoyancy factor. For the limiting case of infinite dilution, f is the same constant as the friction coefficient described for the case of diffusion, so that:

$$S_0 = h^{-1} K_S M^{(2-a)/3} \tag{21}$$

with K_S being obtainable from measurements on linear samples.

Like the diffusion measurement, therefore, measurement of the sedimentation rate in an ultracentrifuge is a method to determine the friction coefficient and hence the branching factor h. In this method it is necessary not only to extrapolate to zero concentration or perform the measurements under theta conditions, but also to extrapolate to a pressure of 1 bar and (for polydisperse samples) to make a correction for the influence of diffusion.[28, 33] For polydisperse samples this method determines the weight-average value of M/f, which, in the non free-draining case, is correlated with the weight-average value of the quotient M/R in accordance with eqn (14).

For non-unperturbed molecules—i.e. molecules having an expanded chain—the hydrodynamic radius is greater. However, the expansion factor α_f is not necessarily equal to the expansion factor of the radius of gyration,[34] $(\alpha_f < \alpha_S)$, and hardly anything is known about the ratio of $(\alpha_f)_{hr}$ to $(\alpha_f)_{lin}$.

If theta conditions are applied, or if a possible difference between $(\alpha_f)_{hr}$ and $(\alpha_f)_{lin}$ is neglected, the hydrodynamic radius can be calculated in a simple way by measurement of the sedimentation rate, and the LCB index h can be determined by making a comparison with data for the linear macromolecule. For fractions of a branched polystyrene under theta conditons Kurata et al.[35] determined h in this way, and calculated the LCB frequency λ from the value thus found, using Kurata and Fukatsu's equation.[8] The λ value they obtained proved to show good correspondence with the value determined from the intrinsic viscosity. Although, according to expectation, S_{br} is in general found to be greater than S_{lin} at equal M,[36-39] there are some reports of the opposite case in the literature $(S_{br} < S_{lin})$.[40, 41] The cause of this is unclear.

For the determination of the branching parameter, h, from the diffusion coefficient it is necessary that the D_0–M relation for linear samples and the (average) molar mass of the polymer concerned be known. To determine h from sedimentation, knowledge of the relation between S_0 and M for linear samples as well as of M is required. As follows from eqns (19) and (21), both M and h can be determined from combined diffusion and sedimentation measurements for monodisperse samples, provided K_D and K_S and also a are known. For polydisperse samples rather complicated averages of M and h are obtained in this way.

3.3. Intrinsic Viscosity

In the non-free-draining case (the normal case with polymer molecules in solution) linear macromolecules satisfy the Flory–Fox equation:[42]

$$[\eta] = 6^{3/2} \, \Phi \, \langle S^2 \rangle^{3/2} M^{-1} \tag{22}$$

in which $[\eta]$ is the intrinsic viscosity, and Φ is a constant.

However, it is also possible to use an equation valid for linear as well as for branched macromolecules:

$$[\eta] = K_v R_v^3 M^{-1} \tag{23}$$

in which K_v is a constant and R_v is the effective hydrodynamic radius. The general principle of $[\eta]$ being proportional to a cubed dimension of the molecule, e.g. the hydrodynamic radius, means that the reduction of the intrinsic viscosity of long-chain branched molecules with respect to the $[\eta]$ of linear molecules having the same molar mass is a measure of the long-chain branching. This reduction of the intrinsic viscosity is generally expressed in the parameter g', which is defined as:

$$g' = [\eta]_{br}/[\eta]_{lin} \tag{24}$$

where $[\eta]_{br}$ is the intrinsic viscosity of the branched molecule and $[\eta]_{lin}$ is the intrinsic viscosity of a linear molecule with the same molar mass, in the same solvent and at the same temperature. $[\eta]_{lin}$ can be calculated from the molar mass M, using the Mark–Houwink equation ($[\eta]_{lin} = KM^a$). Hence, determination of g' requires measurement of $[\eta]$ and M and knowledge of the constants K and a.

If the hydrodynamic radius from eqn (23) should be equal to, or at least proportional to, the hydrodynamic radius defined by the translational friction coefficient (eqn 14), the assumption of Stockmayer and Fixman[11] applies:

$$g' = R_{br}^3/R_{lin}^3 = h^3 \tag{25}$$

If the hydrodynamic radius was proportional to the radius of gyration independently of whether the molecule was branched or linear, it would follow that:[53]

$$g' = \frac{\langle S^2 \rangle_{br}^{3/2}}{\langle S^2 \rangle_{lin}^{3/2}} = g^{3/2} \tag{26}$$

However, it is not to be expected that this condition is satisfied. Because with branched macromolecules the segment density in the centre as compared with the density on the outside is generally greater than in linear molecules, branched molecules have a radius of gyration that is smaller with respect to the hydrodynamic radius than is the case in linear molecules, as a consequence of which g' will be greater than $g^{3/2}$.

In good solvents the effective hydrodynamic radius is enlarged by the factor α_η. This means that the ratio $g' = [\eta]/[\eta]_{\text{lin}}$ is equal to the ratio for the unperturbed state multiplied by the factor $(\alpha_{\eta,\text{br}}/\alpha_{\eta,\text{lin}})^3$. According to experiments by Orofino and Wenger,[43] on star-shaped branched macromolecules, this factor is slightly above unity. The same follows, especially for greater molar masses, from literature data collected by Krozer.[44] Other investigators[45, 46] find hardly any difference between $\alpha_{\eta,\text{br}}$ and $\alpha_{\eta,\text{lin}}$ (for LDPE fractions and star-shaped polystyrene samples), whilst values slightly below unity have also been found for this ratio.[47, 48] Probably one is not far wrong if, in general, one takes $\alpha_{\eta,\text{br}}$ to be equal to $\alpha_{\eta,\text{lin}}$, and hence g' to be independent of the solvent power.

It follows from eqns (14) and (23) that the effective hydrodynamic radius can be determined from both the friction coefficient and the intrinsic viscosity, and that according to eqn (25):

$$g' = h^3$$

There are, indeed, several experimental data that confirm this equality (within a certain accuracy margin).[49-52]

For some well defined branched structures the ratio of the intrinsic viscosities of branched and linear macromolecules has been calculated on the basis of hydrodynamic theories. For some models of star-shaped molecules in the non-free-draining case Zimm and Kilb[54] found a g' value that is very closely approximated by $g^{0.5}$. Consequently, for star-shaped branched macromolecules the functionality of which is not too high, g' might, in general, be put equal to $g^{0.5}$. In the case of macromolecules showing a comb-shaped structure, with side chains that are short as compared with the backbone, the configuration in solution is determined mainly by the (linear) backbone, which means that there is the same ratio of the hydrodynamic radius to the radius of gyration as for linear molecules, and that eqn (26) is equally valid. Therefore, Berry[16] assumes that in the limiting case of combs with relatively short side chains, g' is equal to $g^{1.5}$. For comb-shaped branching with longer side chains and for random branching, an exponent value of between 0·5 and 1·5 may be expected.

In general, therefore, the value of g' may be expressed as:

$$g' = g^b \tag{27}$$

with the exponent b being dependent on the type of branching and at the same time containing the excluded-volume and other effects, both on g and on g'. With a given polymer in a given solvent b may additionally depend on the degree of branching and on the molar mass. Experimentally, values of b have been derived in widely different manners (see, e.g., Table 4.1 in ref. 2); the values found show a wide spread around an average of about unity.[35, 45, 55–63]

Various data found in the literature[50, 51, 64, 65] suggest that for low degrees of branching $b = 0.5$ is a good assumption, whereas for higher degrees of branching it is better to assume $g' = h^3$. For random branching, a comparison of the theoretical formulae of h and g as functions of the number of branch units indicates that h^3 starts off, for very low degrees of LCB, close to $g^{0.5}$, is closely approximated by $g^{0.8}$ for medium-low degrees of branching, and by g for higher degrees of branching.[62, 64] For star-shaped branched macromolecules g' values of between $g^{0.5}$ and h^3 are invariably found.[119]

3.4. Gel Permeation Chromatography

In gel permeation chromatography it is primarily the molecular dimensions which determine the elution volume (V_E) in the separation process. As shown in eqn (23), the hydrodynamic radius is a unique function of the product $[\eta]M$. The universal calibration procedure introduced by Grubisic, Rempp and Benoit,[66] which relates the elution volume to $[\eta]M$, is based on this principle.

A GPC apparatus is normally calibrated with a series of linear polymers. Measurement of a monodisperse linear polymer of the same chemical composition as the polymers used for the calibration directly yields the molar mass. Measurement of a polydisperse linear polymer gives (after correction for axial dispersion) the correct molar mass distribution. With long-chain branched chemically identical samples the data can be treated as if the sample were linear, so that an apparent molar mass M^* and an apparent molar mass distribution $f(M^*)$ will be obtained. M^* and $[\eta]^*$ are the molar mass and the intrinsic viscosity of linear molecules leaving the columns at the same elution volume as the branched molecules with M and $[\eta]$.

For branched molecules, the apparent molar mass, M^*, being related to the molecular dimensions, will be smaller than the real molar mass, M. The ratio M^*/M, therefore, is an index of long-chain branching. For a monodisperse sample:

$$[\eta]^* M^* = K(M^*)^{a+1} \tag{28}$$

and

$$[\eta]M = g'[\eta]_{\text{lin}}M = g'KM^{a+1} \tag{29}$$

K and a are the constants of the Mark–Houwink equation for the linear polymer in the relevant solvent. Application of the universal calibration principle:

$$[\eta]^* M^* = [\eta]M \tag{30}$$

to eqns (28) and (29) leads to:[67]

$$g' = \left(\frac{M^*}{M}\right)^{a+1} \tag{31}$$

This relation between M^*/M and g' is only valid for monodisperse samples. For polydisperse samples the relation between M_w^*/M_w or M_v^*/M_v and g' is more complicated (see section 5).

3.5. Spectroscopic Methods

For characterising branching generally, the determination of end groups is a widely used method. A well known example is the determination of methyl groups in LDPE by infrared spectroscopy. This method, however, does not distinguish between short-chain and long-chain branching.

^{13}C NMR has also proved to be very effective in the determination of branching. Although this method is mainly suited for the determination of short side chains, several investigators have recently shown[68-71, 121] that by using high-resolution apparatus it is possible to determine in LDPE the numbers of the various types of short-chain branches individually, and also, with reasonable accuracy, the number of side chains longer than n-amyl. The results of these experiments generally agree well with the long-chain branching (λ) values calculated from viscometry and GPC or from light scattering and viscometry.

3.6. Determination of LCB from Viscoelastic Properties of Polymers

In addition to the average molar mass and the average molar mass distribution, long-chain branching determines to a high degree the flow behaviour of a polymer melt.[2, 72] If the side chains are not too long the viscosity at zero shear rate (η_0) depends on molecular size (gM_w) only.[73, 74] With greater molecular size, entanglement of the—longer—side chains comes to play a part and η_0 will very rapidly increase, so that it may even become greater than the η_0 of linear samples with the same M_w.[75] At higher shear rates, however, the more non-Newtonian behaviour of the melt leads to lower melt viscosities than exhibited by linear samples.[75, 76]

In connection with the above, it has been proposed to determine the degree of LCB from the viscoelastic properties of the melt, e.g. from the viscosity at zero shear stress, or from the entire flow curve.[77] Since also the melt elasticity depends on the degree of LCB, a parameter such as die swell might also supply information about the long-chain branching of a polymer sample.[78–80] Further, a method has been given by which quantitative LCB determinations can be made from viscoelastic measurements on dilute polymer solutions.[81]

Although in the ways indicated above some—at least qualitative— insight into the presence of long side chains may be gained in a number of cases, a general quantitative determination of the degree of LCB by these methods is still fraught with difficulties.

4. DETERMINATION OF LCB BY COMBINED METHODS (LIGHT SCATTERING, VISCOMETRY, DIFFUSION, SEDIMENTATION, GPC)

As indicated above, the light scattering intensity extrapolated to zero angle depends only on the (mass-average) molar mass, whilst the quantities determined by viscometry, diffusion, sedimentation, and GPC depend also on the hydrodynamic radius. At the present time it is customary in general to equate, or at least to treat as proportional, the hydrodynamic diameters playing a role in these experimental methods, provided the same solvent and the same temperature are used, making allowance for the accuracy of experiments and theory. This implies the validity of the principle of universal calibration (linking viscometry with GPC) and the equality of g' and h^3 (linking viscometry with diffusion and sedimentation). Although originally the use of a combination of light scattering and viscometry was indicated as the way to determine the degree of LCB, in 1962 Moore[82] developed a method to determine g from the sedimentation coefficient and intrinsic viscosity. The quotient $[\eta]/S_0$ indicated by him still contains the expansion factor α, however, whilst the g value he calculated is related to the hydrodynamic radii rather than to the radii of gyration of branched and linear macromolecules. After the appearance of gel permeation chromatography this method was soon used in combination with light scattering and with viscometry for the determination of LCB (see the next section). Tung[83] used a combination of the sedimentation coefficient and GPC data to determine the molar mass distribution and the distribution of the LCB index of polydisperse polymers. Matsuda *et al.*[38, 49] combined diffusion and sedimentation data to determine these distributions.

The use of these combined methods in essence is based on the insight that the physical quantities playing a role all depend on only two molecular parameters, viz. the molar mass and the hydrodynamic radius. For measurements performed in a given solvent and at a given temperature, the same hydrodynamic radius R applies, irrespective of the technique used. Therefore, the following expressions hold, for linear as well as branched molecules, under theta as well as under non-theta conditions:

viscometry: $\quad\quad [\eta] = k_v R^3 / M$ (32)

diffusion: $\quad\quad\quad D_0 = k_D / R$ (33)

sedimentation: $\quad\quad S_0 = k_S M / R$ (34)

GPC: $\quad\quad\quad\quad V_E = f(R) = f'([\eta]M)$ (35)

Further, the following expression holds for the ratio between the hydrodynamic radii of branched and linear molecules:

$$R_{br}/R_{lin} = h = (g')^{1/3} \quad\quad\quad (36)$$

In combination with the Mark–Houwink equation for a linear polymer in the given solvent and at the given temperature:

$$[\eta] = KM^a \quad\quad\quad\quad\quad (37)$$

this leads to the following equations, indicating the dependence of the experimentally determined quantities on the molar mass and the LCB index h (or g'):

light scattering: $\quad M = M$

viscometry: $\quad\quad [\eta] = Kh^3 M^a$

diffusion: $\quad\quad\quad D_0 = K_D h^{-1} M^{-(a+1)/3}$,   or $M_{diff}^{app} = h^{3/(a+1)} M$

sedimentation: $\quad\quad S_0 = K_S h^{-1} M^{(2-a)/3}$,   or $M_{sed}^{app} = h^{-3/(2-a)} M$

GPC: $\quad\quad\quad\quad [\eta]M = Kh^3 M^{a+1}$,   or $M^* = h^{3/(a+1)} M$

$$or \; [\eta]^* = Kh^{3a/(a+1)} M^a \quad (38)$$

$M_{\text{diff}}^{\text{app}}$ is defined by $D_0 = K_D (M_{\text{diff}}^{\text{app}})^{-(a+1)/3}$, the relation between D_0 and M being valid for linear molecules. $M_{\text{sed}}^{\text{app}}$ is defined by $S_0 = K_S (M_{\text{sed}}^{\text{app}})^{(2-a)/3}$, the relation between S_0 and M being valid for linear molecules. Because both D_0 and V_E (GPC) depend on the hydrodynamic radius only, they provide the same information, and $M_{\text{diff}}^{\text{app}}$ is equal to M^*.

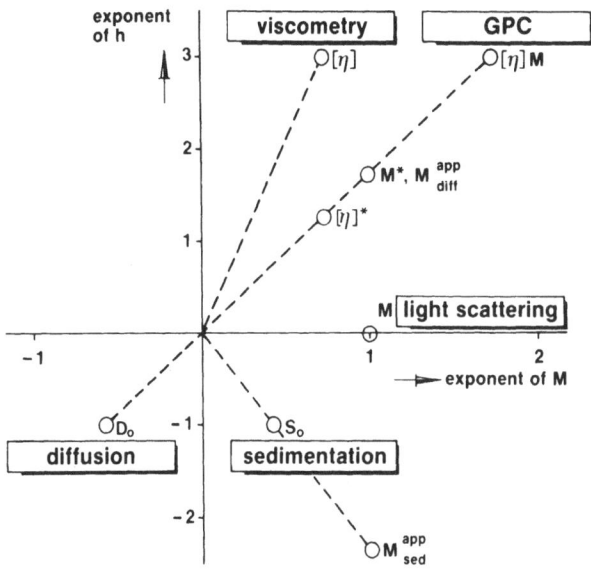

FIG. 1. Dependence on molar mass M and LCB index h of experimental quantities.

Figure 1 shows to what degree the quantities to be determined experimentally depend on the two parameters, molar mass and LCB index. In principle, any pair of the two abovementioned techniques, with the exception of the pair formed by diffusion and GPC, can be used to determine the two parameters M and h or g':

light scattering + viscometry: $\qquad M = M \qquad , h^3 = \dfrac{[\eta]}{KM^a}$

light scattering + diffusion: $\qquad M = M \qquad , h^3 = \left(\dfrac{M_{\text{diff}}^{\text{app}}}{M}\right)^{a+1}$

light scattering + sedimentation: $\qquad M = M \qquad , h^3 = \left(\dfrac{M}{M_{\text{sed}}^{\text{app}}}\right)^{2-a}$

light scattering + GPC:
$$M = M \qquad\qquad , \quad h^3 = \left(\frac{M^*}{M}\right)^{a+1}$$

viscometry + diffusion:
$$M = \frac{K}{[\eta]}(M_{diff}^{app})^{a+1} \qquad , \quad h^3 = \left[\frac{[\eta]}{K}(m_{diff}^{app})^{-a}\right]^{a+1}$$

viscometry + sedimentation:
$$M = \left(\frac{[\eta]}{K}\right)^{1/2}(M_{sed}^{app})^{(2-a)/2} \quad , \quad h^3 = \left[\frac{[\eta]}{K}(M_{sed}^{app})^{-a}\right]^{(2-a)/2}$$

viscometry + GPC:
$$M = \frac{[\eta]M}{[\eta]} = \frac{K(M^*)^{a+1}}{[\eta]} \qquad , \quad h^3 = \left(\frac{[\eta]}{[\eta]^*}\right)^{a+1}$$

diffusion + sedimentation:
$$M = (M_{diff}^{app})^{(a+1)/3}(M_{sed}^{app})^{(2-a)/3}, \; h^3 = \left(\frac{M_{diff}^{app}}{M_{sed}^{app}}\right)^{(a+1)(2-a)/3}$$

sedimentation + GPC:
$$M = (M^*)^{(a+1)/3}(M_{sed}^{app})^{(2-a)/3}, \; h^3 = \left(\frac{M^*}{M_{sed}^{app}}\right)^{(a+1)(2-a)/3} \qquad (39)$$

When the abovementioned assumptions are valid, the correct molar masses and the g' and h values can thus be found for monodisperse samples. With polydisperse samples various average molar masses (the mass-average molar mass in the four cases where light scattering is one of the pair of techniques used) and—generally very complicated—average values of g' are obtained.

Which combination of techniques can best be used depends on the experimental possibilities available and the work involved, on the complexity of the corrections required (e.g. for peak broadening by diffusion in GPC and sedimentation, and for extrapolation in light scattering and sedimentation), and on the accuracy of the final result. From the combinations of two methods, apart from the accuracy of their results, M and g' can be defined more accurately if in the two techniques used LCB has opposite effects on the apparent molar masses.

By means of Tung's combined GPC–sedimentation method[39, 52, 83–85] it is possible to determine the apparent molar mass distribution of a polydisperse sample from GPC, using a calibration for linear samples and making a correction for peak broadening. Likewise it is possible to determine an apparent molar mass distribution from measurement of the sedimentation rate distribution, applying a correction for diffusion and extrapolating to zero concentration and a pressure of 1 bar. Figure 2 shows the integral distribution function of M^* ($F(M^*)$) and that of M_{sed}^{app} ($F(M_{sed}^{app})$). In general, it may be assumed that M^* increases with M_{sed}^{app}, so that equal values of $F(M^*)$ and $F(M_{sed}^{app})$ relate to the same molecules in the distribution. It is then possible to calculate for any value of F the value of M from M^* and M_{sed}^{app} by means of eqn (39), and, using the same equation, to obtain h from the quotient M^*/M_{sed}^{app}. In this way one obtains the absolute molar mass distribution and, at the same time, h as a function of M. The method is equally applicable even if different solvents are used for GPC and sedimentation, assuming h to be independent of the type of solvent used; at worst, the exponents in eqn (39) would be slightly different in the case that

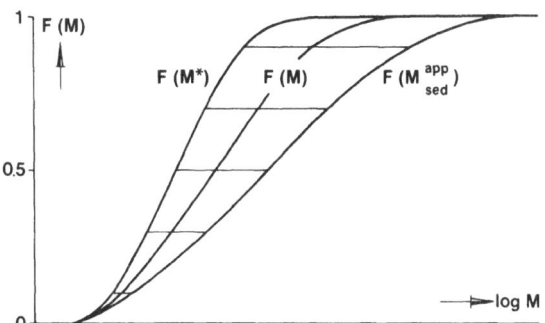

FIG. 2.   Apparent molar mass distribution functions from GPC and sedimentation and $F(M)$ derived from them.

the exponents in the two Mark–Houwink equations had different values. In an analogous way, Matsuda et al.[38, 49] calculated the molar mass distribution and the LCB parameter h as a function of M for a number of polydisperse polymers from diffusion and sedimentation measurements.

5. COMBINATION OF MASS-AVERAGE MOLAR MASS, INTRINSIC VISCOSITY AND GPC DATA

5.1. LCB Indices for Whole Polymers

Let us assume that the mass-average molar mass, M_w, of a long-chain branched polymer sample has been determined by light scattering, the intrinsic viscosity in a given solvent and at a given temperature, $[\eta]$, by means of viscometry, and the overall apparent molar mass distribution, $f(M^*)$, in the same solvent and at the same temperature, by means of GPC.

If the sample is monodisperse, the LCB index g' given by eqn (24) is defined as $[\eta]/[\eta]_{lin}$, with $[\eta]_{lin}$ being derivable from M by means of the Mark–Houwink equation. In a polydisperse sample the various components generally have different g' values. In analogy with eqn (24) the following expression can be used:

$$\langle g' \rangle_{\eta, M_w} = [\eta]/[\eta]_{\text{lin}} \tag{40}$$

which defines the average branching index $\langle g' \rangle_{\eta, M_w}$. In this equation, $[\eta]_{\text{lin}}$ is the intrinsic viscosity of a linear sample having the same molar mass distribution (MMD). Stictly speaking, to calculate $[\eta]_{\text{lin}}$, knowledge of the whole MMD is required, but if the actual distribution can be reasonably approximated by means of a logarithmic normal distribution, knowledge of M_w and of the width of the distribution M_w/M_n will suffice. One can then write:[86]

$$[\eta]_{\text{lin}} = K M_w^a \left(\frac{M_w}{M_n} \right)^{-a(1-a)/2} \tag{41}$$

By analogy with eqn (31), which applies to monodisperse samples, it is possible to define an average branching index g' for a polydisperse sample by postulating:[67, 87]

$$\langle g' \rangle_{M_w, \text{GPC}} = \left(\frac{M_w^*}{M_w} \right)^{a+1} \tag{42}$$

M_w^* being the mass-average apparent molar mass from GPC.

Finally, it is easy to derive (from the universal calibration equation (30) and eqn (31)) that the following relation holds for a monodisperse polymer (see also eqn (39)):

$$g' = \left(\frac{[\eta]}{[\eta]^*} \right)^{a+1} \tag{43}$$

For a polydisperse polymer a third average branching index can now be defined:[67, 87]

$$\langle g' \rangle_{\eta, \text{GPC}} = \left(\frac{[\eta]}{[\eta]^*} \right)^{a+1} \tag{44}$$

with $[\eta]^*$ being the apparent intrinsic viscosity from GPC, i.e. the intrinsic viscosity of a linear sample having the same GP chromatogram as the branched sample. It can easily be calculated from the chromatogram and the Mark–Houwink equation.

This shows that the branching index g' can be derived in three ways from the three experimental data: M_w (from light scattering or by means of the ultracentrifuge), $[\eta]$ (from viscometry), and M^* (M_w^* or $[\eta]^*$) (from gel permeation chromatography) by using different combinations of two of these data. For a monodisperse sample the three methods give the same g' value (assuming the universal calibration principle to have validity). For a polydisperse sample three different g' values can be derived:

$$\langle g' \rangle_{\eta.M_w} = \frac{[\eta]}{[\eta]_{lin}} = \frac{\Sigma f_i g'_i M_i^a}{\Sigma f_i M_i^a} \tag{45}$$

$$\langle g' \rangle_{M_w.GPC} = \left(\frac{M_w^*}{M_w} \right)^{a+1} = \left[\frac{\Sigma f_i (g'_i)^{1/(a+1)} M_i}{\Sigma f_i M_i} \right]^{a+1} \tag{46}$$

$$\langle g' \rangle_{\eta.GPC} = \left(\frac{[\eta]}{[\eta]^*} \right)^{a+1} = \left[\frac{\Sigma f_i g'_i M_i^a}{\Sigma f_i (g'_i)^{a/(a+1)} M_i^a} \right]^{a+1} \tag{47}$$

where f_i is the mass fraction of component i.

In most cases:

$$\langle g' \rangle_{M_w.GPC} < \langle g' \rangle_{\eta.M_w} < \langle g' \rangle_{\eta.GPC} < 1 \tag{48}$$

A combination of broad molar mass distribution, strong long-chain branching, and an increase of LCB (decrease of g') with the molar mass is attended with wide differences between the various g' values. In this case the ratio

$$\frac{\langle g' \rangle_{\eta.GPC}}{\langle g' \rangle_{M_w.GPC}} = \left(\frac{[\eta] M_w}{[\eta]^* M_w^*} \right)^{a+1} \tag{49}$$

is large, too. The dispersion factor expressed in eqn (49) is high, for instance, in some commercial LDPEs having a strongly branched tail of very high molecular weight.

Taking $[\eta]_{lin} = K M_w^a$ and $[\eta]^* = K(M_w^*)^a$ (that is, making no correction for distribution width in the Mark–Houwink equation), one finds from eqns (40), (42) and (44):

$$\langle g' \rangle_{M_w.GPC}^a \cdot \langle g' \rangle_{\eta.GPC} = \langle g' \rangle_{\eta.M_w}^{a+1} \tag{50}$$

This also shows that $\langle g' \rangle_{\eta.M_w}$ has an intermediate value, and that $\langle g' \rangle_{M_w.GPC}$ and $\langle g' \rangle_{\eta.GPC}$ deviate to either side. If the correction for distribution width is applied (see eqn (41)), a slightly more complicated expression is obtained, although the same kind of relation applies.

The practical meaning of the various LCB indices for broad polydisperse samples can also be expressed as follows:

$\langle g' \rangle_{\eta.M_w}$ is an overall LCB index contributed to by all molar masses, although most by the molecules having the highest Ms.

$\langle g' \rangle_{M_w,\text{GPC}}$ is an LCB index affected especially by that part of the sample which has a high molar mass (and is often strongly branched).

$\langle g' \rangle_{\eta,\text{GPC}}$ is an LCB index which is comparatively little affected by such a high-molar-mass part, and which rather provides an impression of the branching belonging to the central section of the molar mass distribution.

5.2. LCB Indices for GPC Fractions

When a sample of a branched polymer is fractionated by means of gel permeation chromatography, a distribution is determined by concentration detection, the M^* of each fraction being derived from the elution volume by means of the GPC calibration. It is also possible to determine the intrinsic viscosity of the eluate by viscosity detection, and its absolute M_w value by light scattering detection, either continuously or batchwise for each syphon charge. For each GPC fraction, the three parameters, M_w, $[\eta]$ and M^*, are known then. From these, the three LCB indices can be calculated by the methods described in the foregoing.

Although these fractions may be regarded as monodisperse in M^* (at least if one neglects the influence of axial dispersion), they are not so in M and g'. In principle, with a given elution volume both branched molecules with a large molar mass, and linear or less pronouncedly branched molecules with a smaller molar mass may be found in the eluate, depending on the degree of homogeneity of the original sample. Each GPC fraction actually contains a range of molecules differing in structure and molar mass. This means that here, too, there is inequality between the various LCB indices. An additional relation between these three g' values is in this case given by eqn (30) which applies to each type of molecule and in which $[\eta]^*$ and M^* are constants for each fraction.

5.2.1. Average Molar Mass of the GPC Fractions

If each GPC fraction is indicated by the index i and each type of molecule in it by the extra index j, eqn (30) reads:

$$[\eta]_{ij} M_{ij} = [\eta]_i^* M_i^* \tag{51}$$

The measured $[\eta]$ of a GPC fraction is:

$$[\eta]_i = \sum_j f_{ij} [\eta]_{ij} = \sum_j (f_{ij}/M_{ij})([\eta]_{ij} M_{ij}) \tag{52}$$

where f_{ij} is the mass fraction of component j in GPC fraction i.

From eqns (51) and (52) it follows that:

$$[\eta]_i = [\eta]_i^* M_i^* \sum_j f_{ij}/M_{ij} = [\eta]_i^* M_i^* (M_n)_i^{-1} \qquad (53)$$

where M_n is the number-average molar mass or:

$$[\eta]_i (M_n)_i = [\eta]_i^* M_i^* = K(M_i^*)^{a+1} \qquad (54)$$

The molar mass derived from M^* and $[\eta]$ through application of the universal calibration equation is therefore the number-average molar mass of the GPC fraction.[55, 88]

5.2.2. Relation Between LCB Indices for GPC Fractions

From eqns (49) and (54) it follows for GPC fraction i:

$$\frac{\langle g' \rangle_{\eta,GPC}}{\langle g' \rangle_{M_w,GPC}} = \left(\frac{(M_w)_i}{(M_n)_i}\right)^{a+1} \qquad (55)$$

In combination with eqn (50) this gives (without correction for width of the distribution in the Mark–Houwink equation):

$$\langle g' \rangle_{M_w,GPC} : \langle g' \rangle_{\eta,M_w} : \langle g' \rangle_{\eta,GPC} = 1 : \left(\frac{M_w}{M_n}\right)_i : \left(\frac{M_w}{M_n}\right)_i^{a+1} \qquad (56)$$

The difference between the three differently derived g' values for a GPC fraction is therefore directly related to the width of the fraction.

Figure 3 shows what actually happens in the GPC fractionation of a broad branched polymer sample. In this figure the two variables, the LCB

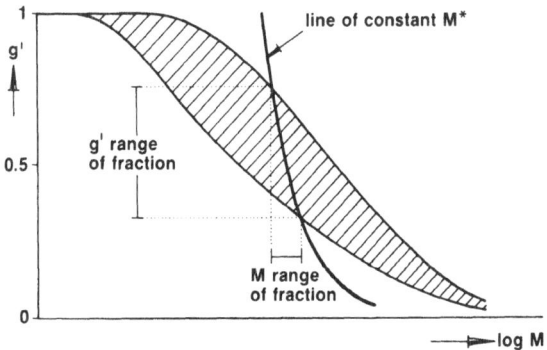

FIG. 3.   Dependence on M and g' of a broad polydisperse long-chain branched sample.

index, g', and the logarithm of the molar mass, log M, of all the sample components have been plotted against each other. In the case of a commercial LDPE sample, for instance, the molecules are in the shaded area (g' decreases with M). At elution volume V_i, the GPC fraction with an apparent molar mass M_i^* leaves the columns. The molecules of this fraction are on that section of a line of constant M^* which is also in the shaded area. This shows that each GPC fraction contains molecules with different M values and also highly different g' values.

6. EFFECT OF SHORT-CHAIN BRANCHING ON g'

In low-density polyethylene, polymer molecules with long-chain branching also show a considerable degree of short-chain branching (SCB). This short-chain branching has less influence on molecular size, but when there is much SCB, the total effect may still become important.[122-124, 57]

If, for simplicity, it is assumed that in each type of molecule the molecular dimensions are not influenced by the short side chains,[89] the size of a macromolecule with molar mass M is the same as that of a molecule with the same long-chain structure but without short side chains. Such a macromolecule has a molar mass $M' = M(1 - S)$, S being the mass fraction of short side chains in the molecule with LCB and SCB.

In linear macromolecules with short side chains, therefore,

$$M^* = M' \text{ and}$$
$$g' = (1 - S)^{a+1} \tag{57}$$

Under theta conditions $a = \frac{1}{2}$ and the ratio of the radii of gyration of linear short chain branched and unbranched macromolecules is:

$$g = M'/M = 1 - S \tag{58}$$

Stockmayer's equation mentioned in reference 1, leads for relatively low values of S to the same result.

For molecules with both long-chain branching and short-chain branching, the effects on the molecular dimensions are cumulative, so:

$$g' = g'_{LCB} g'_{SCB} \tag{59}$$

where g'_{LCB} is the branching index relating to the long-chain branching and g'_{SCB} ($= (1 - S)^{a+1}$) represents the effect of the short-chain branching. For h and g similar equations apply.

In the case of homogeneous incorporation of short-chain branching, this extra factor for the g' values applies to all molecules, and consequently also to the three averages:

$$\langle g' \rangle_{\eta,M_w}, \ \langle g' \rangle_{M_w,\text{GPC}} \text{ and } \langle g' \rangle_{\eta,\text{GPC}}$$

For LDPEs whose mass fraction of short side chains may be up to 10%, g'_{SCB} is approximately 0·85. For samples with a high degree of long-chain branching this factor may be neglected, but in the case of, for instance, copolymers with little LCB and much SCB a correction for the effect of SCB will be necessary.

7. HANDLING OF GEL PERMEATION CHROMATOGRAMS OF LONG-CHAIN BRANCHED POLYMERS

In the case of branched polymers a GP chromatogram can be obtained and, using a calibration for linear polymer, the apparent molar mass (M^*) distribution or, with universal calibration, the $[\eta]M$ distribution can be determined. Since, in general, the degree of long-chain branching increases with the molar mass, the M^* distribution will be narrower than the real MMD.[†]

To derive the real MMD from this apparent distribution, more information is needed than can be obtained from the chromatogram alone. The intrinsic viscosity or the M_w of the whole polymer, or either of these parameters measured in the elution flow in GPC, can provide such additional information (see also ref. 120).

[†] From the apparent molar mass distribution often (particularly with automatic data reduction) apparent molar mass averages are calculated using the formulae:

$$M_n^* = 1/(f_i/M_i^*), \ M_w^* = \Sigma f_i M_i^* \text{ and}$$
$$M_z^* = \Sigma f_i (M_i^*)^2/M_w^*$$

It should be kept in mind, however, that here only M_w^* is the true weight average value of M^*. M_n^* calculated in this way is not the number average value of M^*, unless the factor M^*/M is constant over the whole molar mass range. The correct value of M_n^* is:

$$M_n^* = \frac{\Sigma n_i M_i^*}{\Sigma n_i} = \frac{\Sigma (M_i^*/M_i) f_i}{\Sigma (M_i^*/M_i) f_i/M_i}$$

which can only be calculated if M_i^*/M is known for all M values. A similar reasoning holds for M_z^*.

7.1. GPC with Additional Detection for the Whole Polymer

7.1.1. GPC in Combination with Determination of Intrinsic Viscosity

Some years ago, Drott and Mendelson,[90] and—more or less simultaneously—Kurata *et al.*[91] developed a method for the determination of the molar mass distribution and the degree of long-chain branching from the GP chromatogram and the intrinsic viscosity. For this method the g' vs g relation needs to be known, i.e. the correct exponent b in $g' = g^b$ and the g vs λM or, for polydisperse fractions, the $\langle g \rangle_w$ vs $\langle \lambda M \rangle_w$ relation. To obtain these they made use of the Zimm–Stockmayer equations (4), (5), (6) and (7). The LCB frequency λ is assumed to be constant. The universal calibration in GPC converts the chromatogram into an $[\eta]M$ distribution. By assuming a (provisional) value for λ, it is possible to convert a series of M values into the corresponding values of λM and next to derive the values of g, g', $[\eta]$ and $[\eta]M$. This means that, inversely, the $[\eta]M$ distribution can be converted into an $[\eta]$ distribution from which $[\eta] = \Sigma_i f_i[\eta]_i$ can be calculated where f_i is the mass fraction and $[\eta]_i$ the intrinsic viscosity of GPC fraction i. This calculated $[\eta]$ will not be equal to the experimentally determined $[\eta]$ of the whole polymer, but with an iterative procedure the originally chosen value of λ can be adjusted until $\Sigma_i f_i[\eta]_i$ is equal to the experimentally determined value of $[\eta]$ within the desired margin of accuracy. Both the LCB frequency and the MMD are then known.

Ram and Miltz[92] developed a comparable method, but they did not use the g' vs λM relation according to one of the Zimm–Stockmayer equations with constant parameter λ. Instead, they started from:

$$\ln [\eta] = \ln K + a \ln M, \text{ for } M \leq M_0 \tag{60}$$

and

$$\ln [\eta] = \ln K + a \ln M + b \ln^2 M + c \ln^3 M, \text{ for } M > M_0 \tag{61}$$

where K and a are the constants of the Mark–Houwink equation for the linear polymer, and $c = -b/\ln M_0$. M_0 represents the molar mass at which the long-chain branching begins to have an effect. For any value of M, a randomly selected value of b leads to:

$$[\eta]M = KM^{(1 + a + b \ln M + c \ln^2 M)} \tag{62}$$

This means that for the randomly selected value of b, $[\eta]$ and M can be separately derived from each value of $[\eta]M$ and the $[\eta]M$ distribution can be converted into an $[\eta]$ distribution. By means of an iterative procedure analogous to the method of Drott and Mendelson, the value of b can then be found at which $\Sigma_i f_i[\eta]_i$ is equal to the experimentally determined $[\eta]$ of the whole polymer. With this b value the absolute molar mass distribution and the LCB can be calculated. According to Wild $et\ al.$,[93] the Drott–Mendelson method and the Ram–Miltz method are equivalent as regards accuracy of results. As indicated in section 5.2.1 it is the M_n of the GPC fractions that follows from the eluation volume and, therefore, must be used both in Drott and Mendelson's and in Ram and Miltz's method.[126]

7.1.2. GPC Combined with M_w

As a second parameter additional to the GP chromatogram it is also possible to take the M_w of the whole polymer. In that case a kind of Drott–Mendelson analysis has to be applied. However, for the randomly selected value of λ not the $[\eta]$ distribution but the M distribution is derived from the $[\eta]M$ distribution, and the M_w calculated from it is compared with the M_w measured by means of light scattering. An iterative procedure then leads to the correct λ and the correct MMD. It is also possible to use the method of Ram and Miltz and iterate to arrive at the value of $\Sigma\ f_iM_i$ which is equal to the measured M_w of the sample.

An accurate determination of M_w, for instance by means of light scattering, requires a greater effort than a determination of $[\eta]$. Thus, it is necessary that the sample is cleaned in such a way that an M_w value is obtained that is not influenced by microgel or other extraneous material. On the other hand, in averaging over all molecules, the high-molar-mass and strongly branched components have a greater influence on M_w than on $[\eta]$, which means that the branching density can be more accurately established with a combination of M_w and GPC. If the condition of constancy of λ cannot be fulfilled completely, a combination of M_w and GPC tends to establish the average value of λ in the high-molar-mass part of the MMD, while a combination of GPC and $[\eta]$ tends to establish the average λ in the more central part of the MMD.

Experience shows that in polyethylenes, minor quantities of very high-molecular and strongly branched material do not have an appreciable effect on the processing and product characteristics in spite of a strong effect on M_w from light scattering. This implies that in these cases the combination of GPC and viscometry (and $\langle g'\rangle_{\eta,GPC}$ of the whole sample) better characterises the total sample than the combination of GPC and M_w (leading to $\langle g'\rangle_{M_w,GPC}$) does.

7.2. GPC with Additional Detection in the Eluate

The other methods are based on measurement of the additional datum ($[\eta]$ or M_w) of all GPC fractions. These methods are not based on the assumption of constant branching density throughout the MMD, but permit the determination of LCB index g' and the LCB frequency λ as functions of the molar mass. They require more measurements, but provide a good deal more information and are generally applicable.

7.2.1. GPC in Combination with Viscosity Measurement of the Eluate

Several investigators succeeded in measuring the viscosity of the GPC eluate. Experimentally, this can be effected in two ways. In the first method[94-96] a conventional capillary viscometer, e.g. one of the Ubbelohde type, is placed near the GPC apparatus. At regular intervals the syphon contents are transferred into the viscometer reservoir and measured. The normal concentration detector of the gel permeation chromatograph gives the concentration of the eluate sample examined. A limited number of values of the intrinsic viscosity are thus obtained, each of which is an average for the contents of a count. Modern high speed GPC apparatus with small elution volumes makes use of a batch viscometer very difficult.

The other procedure is based on measurement of the difference in pressure between the inlet and the outlet of a capillary placed in the eluate stream.[97-100] This method requires a highly constant flow rate and a highly constant temperature, as well as very accurate measurement of the pressure difference. In combination with the concentration detection, this gives continuous information on the intrinsic viscosity. Since the concentration of the eluate is very low, extrapolation to zero concentration can be dispensed with.

The elaboration of the data from GPC and the intrinsic viscosity measurements may be disturbed by interference from the axial dispersion in the GPC.[101, 102] Molecules of a size which corresponds to a certain elution volume V_E leave the columns not only at that elution volume, but also at elution volumes a little below or above V_E. Mostly the effect of this axial dispersion on the chromatogram is not very strong. Only in the case of very narrow peaks is correction for this peak-broadening effect necessary.

The effect of the axial dispersion on the relation between $[\eta]$ and M^* is much stronger, however. A fraction on the low-molar-mass side of the peak contains many molecules of a size larger than that which corresponds to the elution volume. When relating the measured $[\eta]$ to the $[\eta]^*$ corresponding to this elution volume in accordance with eqn. (43), one will find too high a value for g'. The calculated g' may even exceed unity then. On the high-molar-mass side of the peak of the chromatogram, by contrast, many molecules have smaller dimensions than those corresponding to V_E.

When relating the measured value of $[\eta]$ to the $[\eta]^*$ value which corresponds to V_E, one finds a value of g' which is too low (see Fig. 4).

To avoid these errors in the determination of g', it is necessary to make a correction for axial dispersion. In the literature a number of correction methods are indicated, most of which start from a numerical solution of the integral equation of Tung.[103-105] In the case under review, however, it is necessary not only to reduce the GP chromatogram to the correct distribution over the elution volumes, but also to correct the experimentally obtained distribution of $[\eta]c$ over V_E.[106] Dividing the so-corrected value of

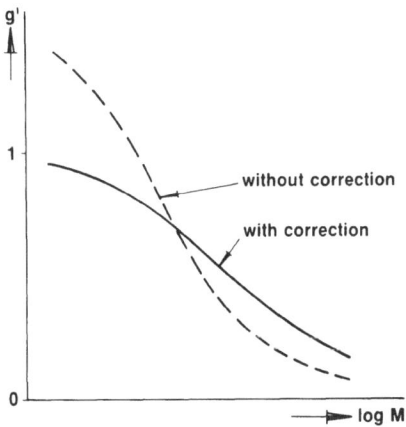

FIG. 4. g' determined by GPC and viscometry of the fractions.

$[\eta]c$ by the corrected c value for each V_E gives the correct intrinsic viscosities. Next, by means of eqn (54) the molar mass over the entire MMD can be derived from M^* and $[\eta]$. Although actually one determines in this way the M_n of the GPC fraction, which is assumed to be infinitely narrow in M^*, it is allowable to assume this fraction to be very narrow in M as well and to calculate from these M values the correct MMD and also the LCB index g' as a function of M.

7.2.2. GPC in Combination with the Light-Scattering Measurement on the Eluate

Also the intensity of the scattered light can be determined as a function of the elution volume. This can be performed either by collecting the eluate at regular intervals in a light-scattering cell for separate intensity measurements, or by passing the eluate through the cell of a light-scattering photometer.[107–110, 125] As the eluate passes through the GPC columns, additional dust from the column charge may be entrained; it is also possible that the columns retain dust already present in the eluate. In any case, thorough filtration of the eluate before it enters the light-scattering meter is required.

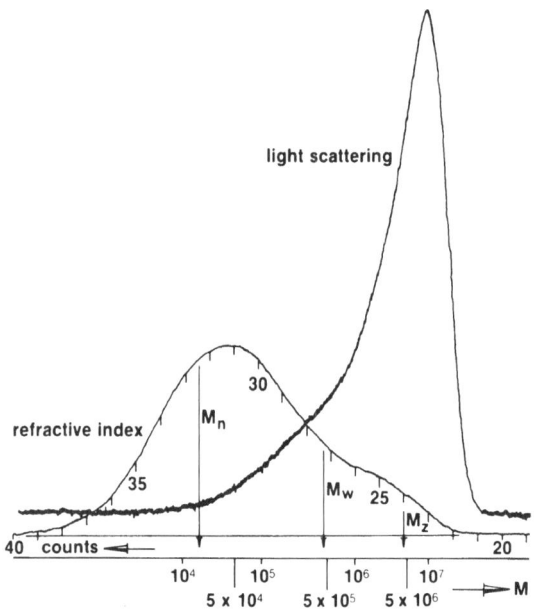

FIG. 5.   Concentration chromatogram and light-scattering chromatogram of a commercial LDPE sample.

Figure 5 shows a combination of the normal chromatogram (concentration signal from differential refractometer) and the light-scattering signal (from low-angle laser light-scattering photometer) for a GPC run with a commercial LDPE sample in 1,2,4-trichlorobenzene.

Combination of the light scattering intensity with the concentration, determined, e.g., by means of a differential refractometer, yields the mass-average molar mass of the fractions. Assuming the GPC fractions to be very narrow, this enables the absolute MMD to be derived, also in the case of long-chain branched samples. With a branched polymer, the calculation of the degree of long-chain branching from the combination of the M_w thus determined and the elution volume (so M^*) of each fraction requires a correction for axial dispersion, analogous to that applied in the case of GPC combined with $[\eta]$ measurements. If such a correction is not applied, the derived g' values will be incorrect, in a sense contrary to the deviations occurring in the uncorrected calculation of g' from combined GPC and viscosity data.

Referring to measurements on a commercial LDPE sample, Table 1 shows the molecular data for a series of GPC fractions determined from GPC calibration, light scattering, viscometry and concentration detection, without correction for axial dispersion. It also contains the g' values calculated from M_w and $[\eta]$, from M_w and M_w^* and from $[\eta]$ and $[\eta]^*$. Figure 6 shows these uncorrected g' values as functions of M. Apart from a very small deviation due to the width of the distribution, the $\langle g' \rangle_{\eta.M_w}$ data are

TABLE 1

MOLECULAR DATA FOR A SERIES OF GPC FRACTIONS OF A COMMERCIAL LDPE, CALCULATED WITHOUT CORRECTION FOR AXIAL DISPERSION

Fraction No.	M_w^* (kg/mol)	$[\eta]^*$ (dl/g)	M_w (kg/mol)	$[\eta]_{TCB}^{135°}$ (dl/g)	$\langle g' \rangle_{\eta,M_w}$	$\langle g' \rangle_{M_w,GPC}$	$\langle g' \rangle_{\eta.GPC}$
1	2 800	18·10	9 000	4·4	0·11	0·13	0·09
2	810	7·3	1 700	2·1	0·18	0·28	0·12
3	240	3·0	580	1·53	0·28	0·22	0·31
4	86	1·45	275	1·08	0·34	0·13	0·60
5	40	0·83	85	0·70	0·52	0·27	0·75
6	17	0·45		0·43			0·93
7	7	0·23		0·28			1·4
8	3	0·13		0·22			2·5

correct values. $\langle g' \rangle_{M_w.GPC}$ and $\langle g' \rangle_{\eta.GPC}$, however, are incorrect, especially at the sides of the molar mass distribution. Use of corrected values of the molecular data brings the three curves in Fig. 6 together. In principle, the differences between the three values can be used to determine the dispersion parameter for a given combination of polymer solution and GPC apparatus.

Because of the very low polymer concentration in the GPC eluate, extrapolation to zero concentration is not necessary, either for viscosity or for light-scattering measurements. If desired, the concentration dependence can be allowed for by introduction of an estimated value of the Huggins constant or the second virial coefficient, respectively. On the other hand, the very low concentration entails the necessity of highly accurate measurement in order to obtain reliable results.

Insofar as viscosity measurements are concerned, it seems that the method is suitable for routine purposes.[55, 101, 102, 106, 111-113] As regards on-line light-scattering measurement, substantial progress has been made in the past few years,[114-118] although, on account of the low concentration, the accuracy on the low-molar-mass side of the MMD (M between 10^4 and 10^5 g/mol) can still be a problem. However, in this range of the MMD, the degree of LCB is often very low in practice, so that reasonably reliable results can be obtained with the GPC calibration for linear polymers and an

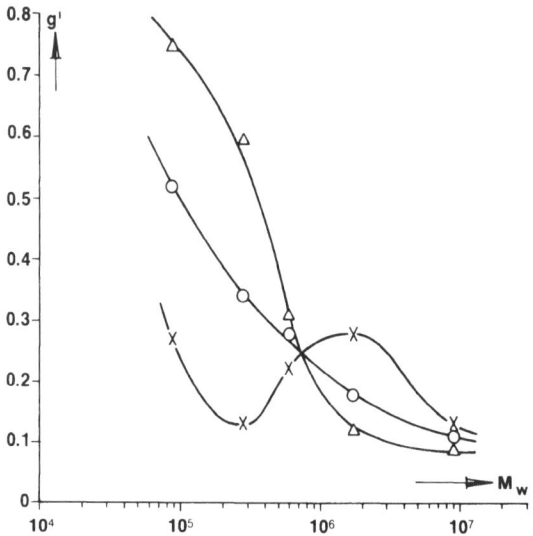

FIG. 6. Influence of axial dispersion; g' determined from uncorrected molecular data for a series of GPC fractions of a commercial LDPE sample. ——o—— g' from M_w and $[\eta]$ of the fractions; ——×—— g' from M_w and M_w^* of the fractions; ——Δ—— g' from $[\eta]$ and $[\eta]^*$ of the fractions.

estimated extrapolation of g' to 1. The Chromatix low-angle laser light scattering photometers KMX-6 and CMX-100 are suitable instruments for use in GPC.[109, 118]

For complete characterisation of polymers with long-chain branching, GPC with on-line viscosity and light-scattering detection is an almost ideal method. It uses the gel permeation chromatograph as a fractionation apparatus only and makes calibration superfluous. Moreover, detailed correction for axial dispersion is not necessary any more; approximate estimation of the polydispersity of the eluate suffices.

8. REDUCTION OF THE LCB INDICES h AND g' TO LCB FREQUENCY λ

Having determined the LCB index g' or h for an LCB sample, whether or not as a function of the molar mass, reduction of this parameter to the number of long side-chains is a procedure which is beset with many uncertainties (see sections 2 and 3 of this chapter). In the first place, it requires knowledge of the branching structure. But even if the type of structure is known, there is the problem that for many structures the relation of λ to g' and M is not fully established, even in the unperturbed state. Furthermore, this relation may be influenced—in a way not satisfactorily investigated so far—by the polymer–solvent interaction and by the degree of draining, which is not fully known. For the LCB index h the same holds, although to a lesser extent.

If b is known, or at least a certain value of b has been assumed, it is possible—in the case of branched samples—first to reduce g' to g, and next, with the formulae shown in the foregoing, to calculate λ. For randomly branched polymers this is mostly done with the equations derived by Zimm and Stockmayer, for whole samples with those applying to polydisperse, and for GPC fractions with those applying to monodisperse samples. In the case of long-chain branching induced by radical polymerisation (through chain transfer), it is obvious to use formulae for a functionality f of 3. In the case of branching caused by copolymerisation with dienes, the formulae for $f = 4$ should be used. Starting from the LCB index h, determined directly or by assuming that $g' = h^3$, λ is calculated with the appropriate formulae of Kurata and Fukatsu or of Stockmayer and Fixman.

The experimentally simplest and most reliable method is the combination of GPC and viscometry. In processing data from branched samples of which the GP chromatogram and $[\eta]$ are known, it is possible—if $[\eta]$ has been measured for the separate fractions—to calculate g', and subsequently λ, for each fraction from $[\eta]$ and $[\eta]^*$. If only the $[\eta]$ of the whole polymer is known, a direct determination of λ is not possible. It is best then to use the Drott–Mendelson method, which automatically yields the parameter λ.

It should be kept in mind, of course, that all λ values arrived at by any of the abovementioned methods are more or less of an approximate nature. For comparison of samples which do not differ very much, λ is a suitable parameter, though, and more significant than g', for instance, as λ is not by nature highly dependent on the molar mass.

34   TH. G. SCHOLTE

REFERENCES

1. BILLMEYER, F. W., *J. Amer. Chem. Soc.*, 1953, **75**, 6118.
2. SMALL, P. A., *Adv. Polym. Sci.*, 1975, **18**, 1.
3. YAMAKAWA, H., *Modern Theory of Polymer Solutions*, 1971, Harper and Row, New York.
4. KRAMERS, H. A, *J. Chem. Phys.*, 1946, **14**, 415.
5. ZIMM, B. H. and STOCKMAYER, W. H., *J. Chem. Phys.*, 1949, **17**, 1301.
6. See Ref. 3, page 48.
7. OROFINO, T. A., *Polymer*, 1961, **2**, 305.
8. KURATA, M. and FUKATSU, M., *J. Chem. Phys.*, 1964, **41**, 2934.
9. FLORY, P. J., *Chem. Rev.*, 1946, **39**, 137.
10. STOCKMAYER, W. H., *J. Chem. Phys.*, 1943, **11**, 45.
11. STOCKMAYER, W. H. and FIXMAN, M., *Ann. N.Y. Acad. Sci.*, 1953, **57**, 334.
12. PTITSYN, O. B., *Zh. Tekhn. Fiz.*, 1959, **29**, 75.
13. FLORY, P. J., *Principles of Polymer Chemistry*, 1953, Cornell University Press, Ithaca, NY.
14. BERRY, G. C. and OROFINO, T. A., *J. Chem. Phys.*, 1964, **40**, 1614.
15. BERRY, G. C. and CASASSA, E. F., *J. Polym. Sci.*, *D*, 1970, **4**, 1.
16. BERRY, G. C., *J. Polym. Sci.*, *A-2*, 1971, **9**, 687.
17. MAZUR, J. and McCRACKIN, F., *Macromolecules*, 1977, **10**, 326.
18. LEEMPUT, R. V. and MEUNIER, J., *Makromol. Chem.*, 1971, **147**, 191.
19. KUHN, R., KRÖMER, H. and ROSSMANITH, G., *Angew. Makromol. Chem.*, 1974, **40/41**, 361.
20. KRIGBAUM, W. R. and TREMENTOZZI, Q. A., *J. Polym. Sci.*, 1958, **28**, 295.
21. CASASSA, E. F., *J. Chem. Phys.*, 1962, **37**, 2176.
22. CASASSA, E. F., *J. Chem. Phys.*, 1964, **41**, 3213.
23. OTOCKA, E. P., ROE, R. J., HELLMAN, M. Y. and MUGLIA, P. M., *Macromolecules*, 1971, **4**, 507.
24. GIANOTTI, G., CICUTA, A. and ROMANINI, D., *Polymer*, 1980, **21**, 1087.
25. KRATOCHVÍL, P., in: *Light Scattering from Polymer Solutions*, M. B. Huglin, Ed., 1972, Academic Press, London and New York, p. 333.
26. KRATOHVIL, J. P., *Anal. Chem.*, 1966, **38**, 517R.
27. LANGE, H., *Kolloid Z. u. Z. Polymere*, 1970, **240**, 747.
28. FUJITA, H., *Foundations of Ultracentrifugal Analysis*, 1975, John Wiley & Sons, New York.
29. KIRKWOOD, J. G. and RISEMAN, J., *J. Chem. Phys.*, 1948, **16**, 565.
30. BERNE, B. J. and PECORA, R., *Dynamic Light Scattering*, 1976, Wiley–Interscience, New York.
31. BURCHARD, W. and SCHMIDT, M., *Polym. Prepr., Am. Chem. Soc., Div. Polym. Chem.*, 1979, **20** (2), 164.
32. BERRY, G. C., *Polym. Prepr., Am. Chem. Soc., Div. Polym. Chem.*, 1979, **20** (2), 168.
33. SCHACHMAN, H. K., *Ultracentrifugation in Biochemistry*, 1959, Academic Press, London and New York.
34. STOCKMAYER, W. H. and ALBRECHT, A. C., *J. Polym. Sci.*, 1958, **32**, 215.
35. KURATA, M., ABE, M., IWAMA, M. and MATSUSHIMA, M., *Polymer J.*, 1972, **6**, 729.

36. PODDUBNYI, I. YA. and GRECHANOVSKII, V. A., *Vysokomol. Soed.*, 1964, **6**, 64.
37. ROOVERS, J. E. L. and BYWATER, S., *Macromolecules*, 1972, **5**, 384.
38. MATSUDA, H., YAMADA, I. and KUROIWA, S., *Polymer J.*, 1976, **8**, 415.
39. TUNG, L. H. and KNIGHT, G. W., *J. Polym. Sci.*, *A-2*, 1969, **7**, 1623.
40. BLUMSTEIN, A. and BILLMEYER, F. W., *J. Polym. Sci.*, *A-2*, 1966, **4**, 465.
41. WALES, M. and COLL, H., *Ann. N.Y. Acad. Sci.*, 1969, **164**, 102.
42. FLORY, P. J. and FOX, T. G., *J. Am. Chem. Soc.*, 1951, **73**, 1904.
43. OROFINO, T. A. and WENGER, F., *J. Phys. Chem.*, 1963, **67**, 566.
44. KROZER, S., *Makromol. Chem.*, 1974, **175**, 1905.
45. HAMA, T., YAMAGUCHI, K. and SUZUKI, T., *Makromol. Chem.*, 1972, **155**, 283.
46. HERZ, J., HERT, M. and STRAZIELLE, C., *Makromol. Chem.*, 1972, **160**, 213.
47. ROOVERS, J. E. L. and BYWATER, S., *Macromolecules*, 1974, **7**, 443.
48. HADJICHRISTIDIS, N. and ROOVERS, J. E. L., *J. Polym. Sci.*, *Poly. Phys. Ed.*, 1974, **12**, 2521.
49. MATSUDA, H., YAMADA, I., OKABE, M. and KUROIWA, S., *Polymer J.*, 1977, **9**, 527.
50. BERRY, G. C., HOBBS, L. M. and LONG, V. C., *Polymer*, 1964, **5**, 31.
51. AMBLER, M. R., *J. Appl. Polym. Sci.*, 1977, **21**, 1655.
52. DIETZ, R. and FRANCIS, M. A., *Polymer*, 1979, **20**, 450.
53. THURMOND, C. D. and ZIMM, B. H., *J. Polym. Sci.*, 1952, **8**, 477.
54. ZIMM, B. H. and KILB, R. W., *J. Polym. Sci.*, 1959, **37**, 19.
55. FOSTER, G. N., MACRURY, T. B. and HAMIELEC, A. E., in: *Liquid Chromatography of Polymers and Related Materials—II*, J. Cazes and X. Delamare, Eds, 1980, Marcel Dekker, New York, p. 143.
56. MOORE, W. R. A. D. and MILLNS, W., *Brit. Poly. J.*, 1969, **1**, 81.
57. VÖLKER, H. and LUIG, F.-J., *Angew. Makromol. Chem.*, 1970, **12**, 43.
58. COTE, J. A. and SHIDA, M., *J. Polym. Sci.*, *A-2*, 1971, **9**, 421.
59. MRKVIČKOVÁ-VACULOVÁ, L. and KRATOCHVÍL, P., *Coll. Czechoslov. Chem. Commun.*, 1972, **37**, 2015 and 2029.
60. HOFFMANN, M. and KUHN, R., *Makromol. Chem.*, 1973, **174**, 149.
61. CASPER, R., BISKUP, U., LANGE, H. and POHL, U., *Makromol. Chem.*, 1976, **177**, 1111.
62. BODHANECKÝ, M., *Macromolecules*, 1977, **10**, 971.
63. POPOV, G., GEHRKE, K. and ULBRICHT, J., *Plaste u. Kautschuk*, 1974, **21**, 515.
64. NAGASUBRAMANIAN, K., SAITO, O., and GRAESSLEY, W. W., *J. Polym. Sci.*, *A-2*, 1969, **7**, 1955.
65. SCHRÖDER, E., *Plaste u. Kautschuk*, 1973, **20**, 241.
66. GRUBISIC, Z., REMPP, P. and BENOIT, H., *J. Polym. Sci.*, *B*, 1967, **5**, 753.
67. PRECHNER, R., PANARIS, R. and BENOIT, H., *Makromol. Chem.*, 1972, **156**, 39.
68. BOVEY, F. A., SCHILLING, F. C., MCCRACKIN, F. L. and WAGNER, H. L., *Macromolecules*, 1976, **9**, 76.
69. AXELSON, D. E., LEVY, G. C. and MANDELKERN, L., *Macromolecules*, 1979, **12**, 41.
70. FOSTER, G. N., *Polym. Prepr.*, *Am. Chem. Soc.*, *Div. Polym. Chem.*, 1979, **20** (2), 463.

71. HOFFMANN, W.-D., ECKHARDT, G., BRAUER, E. and KELLER, F., *Acta Polym.*, 1980, **31**, 233.
72. FERRY, J. D., *Viscoelastic Properties of Polymers*, 3rd Edition, 1980, John Wiley & Sons, New York.
73. BERRY, G. C. and FOX, T. G., *Adv. Polym. Sci.*, 1968, **5**, 261.
74. PEDERSEN, S. and RAM, A., *Polym. Eng. Sci.*, 1978, **18**, 990.
75. KRAUS, G. and GRUVER, J. T., *J. Polym. Sci., A*, 1965, **3**, 105.
76. MENDELSON, R. A., BOWLES, W. A. and FINGER, F. L., *J. Polym. Sci.*, A–2, 1970, **8**, 105.
77. GHIJSELS, A. and MIERAS, H. J. M. A., *J. Polym. Sci., Poly. Phys. Ed.*, 1973, **11**, 1849.
78. MENDELSON, R. A. and FINGER, F. L., *J. Appl. Polym. Sci.*, 1973, **17**, 797.
79. WILD, L., RANGANATH, R. and KNOBELOCH, D. C., *Polym. Eng. Sci.*, 1976, **16**, 811.
80. STARCK, P. and LINDBERG, J. J., *Angew. Makromol. Chem.*, 1979, **75**, 1.
81. MITSUDA, Y., SCHRAG, J. L. and FERRY, J. D., *J. Appl. Polym. Sci.*, 1974, **18**, 193.
82. MOORE, L. D., GREEAR, G. R. and SHARP, J. O., *J. Polym. Sci.*, 1962, **59**, 339.
83. TUNG, L. H., *J. Polym. Sci.*, A–2, 1969, **7**, 47.
84. TUNG, L. H., *J. Polym. Sci.*, A–2, 1971, **9**, 759.
85. DIETZ, R., *J. Appl. Polym. Sci.*, 1980, **25**, 951.
86. KONINGSVELD, R. and TUIJNMAN, C. A. F., *Makromol. Chem.*, 1960, **38**, 39.
87. SCHOLTE, TH.G. and MEIJERINK, N. L. J., *Brit. Polym. J.*, 1977, **9**, 133.
88. HAMIELEC, A. E. and OUANO, A. C., *J. Liq. Chromat.*, 1978, **1**, 111.
89. TONELLI, A. E., *J. Am. Chem. Soc.*, 1972, **94**, 2972.
90. DROTT, E. E. and MENDELSON, R. A., *J. Polym. Sci.*, A–2, 1970, **8**, 1361 and 1373.
91. KURATA, M., OKAMOTO, H., IWAMA, M., ABE, M. and HOMMA, T., *Polymer J.*, 1972, **3**, 739.
92. RAM, A. and MILTZ, J., *J. Appl. Polym. Sci.*, 1971, **15**, 2639.
93. WILD, L., RANGANATH, R. and BARLOW, A., *J. Appl. Polym. Sci.*, 1977, **21**, 3331.
94. MEYERHOFF, G., *Makromol. Chem.*, 1968, **118**, 265; *Separation Sci.*, 1971, **6**, 239.
95. GOEDHART, D. and OPSCHOOR, A., *J. Polym. Sci.*, A–2, 1970, **8**, 1227.
96. GRUBISIC-GALLOT, Z., PICOT, M., GRAMAIN, PH. and BENOIT, H., *J. Appl. Polym. Sci.*, 1972, **16**, 2931.
97. OUANO, A. C., *J. Polym. Sci., A–1*, 1972, **10**, 2169.
98. OUANO, A. C., *J. Polym. Sci., Symp.*, 1973, **43**, 299.
99. LESEC, J. and QUIVORON, C., *Analusis*, 1976, **4**, 456.
100. LETOT, L., LESEC, J. and QUIVERON, C., *J. Liq. Chromat.*, 1980, **3**, 427.
101. MARAIS, L., GALLOT, Z. and BENOIT, H., *Analusis*, 1976, **4**, 443.
102. PARK, W. S. and GRAESSLEY, W. W., *J. Polym. Sci., Polym. Phys Ed.*, 1977, **15**, 71.
103. TUNG, L. H., *J. Appl. Polym. Sci.*, 1966, **10**, 375.
104. FRIIS, N. and HAMIELEC, A. E., *Adv. Chromat.*, 1975, **13**, 41.
105. HAMIELEC, A. E., EDERER, H. J. and EBERT, K. H., *J. Liq. Chromat.*, 1981, **4**, 1697.

106. SERVOTTE, A. and DE BRUILLE, R., *Makromol. Chem.*, 1975, **176**, 203.
107. CANTOW, H. J., SEIFERT, E. and KUHN, R., *Chem. Ing. Techn.*, 1966, **38**, 1032.
108. OUANO, A. C. and KAYE, W., *J. Polym. Sci.*, 1974, **12**, 1151.
109. OUANO, A. C., *J. Chromat.*, 1976, **118**, 303.
110. MILLAUD, B. and STRAZIELLE, C., *Makromol. Chem.*, 1979, **180**, 441.
111. NAKANO, S. and GOTO, Y., *J. Appl. Polym. Sci.*, 1976, **20**, 3313.
112. CONSTANTIN, D., *Europ. Polym. J.*, 1977, **13**, 907.
113. SCHEINERT, W., *Angew. Makromol. Chem.*, 1977, **63**, 117.
114. HAMIELEC, A. E., OUANO, A. C. and NEBENZAHL, L. L., *J. Liq. Chromat.*, 1978, **1**, 527.
115. MACRURY, T. B. and MCCONNELL, M. L., *J. Appl. Polym. Sci.*, 1979, **24**, 651.
116. KATO, T., KANDA, A., TAKAHASHI, A., NODA, I., MAKI, S. and NAGASAWA, M., *Polymer J.*, 1979, **11**, 575.
117. AXELSON, D. E. and KNAPP, W. C., *J. Appl. Polym. Sci.*, 1980, **25**, 119.
118. JORDAN, R. C., *J. Liq. Chromat.*, 1980, **3**, 439.
119. BAUER, B. J. and FETTERS, L. J., *Rubber Chem. Tech.*, 1978, **51**, 3.
120. DROTT, E. E., in: *Liquid Chromatography of Polymers and Related Materials*, J. Cazes, Ed., 1977, Marcel Dekker, New York and Basel, p. 161.
121. RANDALL, J. C., in: *Polymer Characterization by ESR and NMR*, A. E. Woodward and F. A. Bovey, Eds, ACS Symposium Series 142, 1980, American Chemical Society, Washington, DC, p. 93.
122. TREMENTOZZI, Q. A., *J. Polym. Sci.*, 1956, **22**, 187.
123. GUILLET, J. E., *J. Polym. Sci.*, A, 1963, **1**, 2869.
124. SCHRÖDER, E. and WINKLER, E., *Plaste u. Kautschuk*, 1974, **21**, 269.
125. JORDAN, R. C. and MCCONNELL, M. L., in: *Size Exclusion Chromatography*, T. Provder, Ed, ACS Symposium Series 138, 1980, American Chemical Society, Washington, D.C., p. 107.
126. FOSTER, G. N., HAMIELEC, A. E. and MACRURY, T. B., in: *Size Exclusion Chromatography*, T. Provder, Ed., ACS Symposium Series 138, 1980, American Chemical Society, Washington, D.C., p. 131.

Chapter 2

APPLICATIONS OF THERMAL ANALYSIS TO THE STUDY OF POLYMER BLENDS

J. R. FRIED

Department of Chemical and Nuclear Engineering, University of Cincinnati, Ohio, USA

SUMMARY

Thermal analysis provides a rapid means of assessing the compatibility of small samples of polymer blends. In this chapter, the application and practical limitations of thermal analysis in the study of polymer blends are explored. Thermodynamic principles of blend miscibility are presented and the usefulness of thermal methods are discussed in relation to other techniques frequently used to characterise polymer blends. Examples of technically significant polymer blends are given and a detailed discussion of quantitative thermal analysis of both compatible and phase separated blends is presented with attention to areas of current controversy.

1. INTRODUCTION

Blending two or more polymers is a versatile method of obtaining new materials with improved properties. Such blends have important commercial applications and have been the subject of several recent reviews[1-3] and monographs.[4,5] Depending upon the thermodynamics of polymer–polymer interactions and the kinetics of the mixing process, polymer blends may be homogeneous or phase separated materials. Analogous to plasticisation by low molecular weight additives, homogeneous blending with a polymer of a low glass transition temperature (T_g) may be used to reduce the melt viscosity of a polymer with an otherwise high T_g and poor thermal stability. Important commercial examples of such applications include a variety of polymeric plasticisers for poly(vinyl chloride) (PVC) and blends of poly(2,6-dimethyl-1,4-phenylene oxide), abbreviated PMMPO, with polystyrene (PS). In the case of PMMPO/PS blends, blending has the additional advantage of reducing the cost of an expensive engineering thermoplastic (PMMPO) through addition of a low priced commodity thermoplastic.

Heterogeneous blending may be used to increase the impact resistance of brittle plastics or to improve processibility and surface finish. Examples of impact modification include rubber-filled thermoplastics such as ABS resins and high impact polystyrene (HIPS). In these cases of impact modified resins, good mechanical integrity is provided through graft copolymerisation between polymers in the dispersed rubbery phase and glassy matrix. Polymeric lubricants include polyolefins which are widely used as additives for PVC and other important thermoplastics.

Commercial blend resins may include both compatible (homogeneous) and incompatible polymer components to achieve a combination of different property modifications. For example, Noryl[R] resins[6] are commercial blends of PMMPO with HIPS and low density polyethylene (LDPE) added as a lubricating agent.[7] In this case, PMMPO and PS (the latter being the compatible glassy component of HIPS) form the homogeneous phase whose T_g or heat distortion temperature can be varied over a range of nearly 100°C by combining different ratios of PMMPO and HIPS. The dispersed rubbery phase of HIPS enables improvement in the impact resistance of the resin while LDPE and other additives serve to improve its processibility.

1.1. Thermodynamics of Polymer–Polymer Miscibility

Whether or not two polymers may be thermodynamically miscible is governed by the sign of the Gibbs free energy of mixing (ΔG_M). The free energy is related to temperature (T) and to the enthalpy (ΔH_M) and entropy (ΔS_M) of mixing through the expression

$$\Delta G_M = \Delta H_M - T\Delta S_M \tag{1}$$

If ΔG_M is positive over the entire composition range at a given temperature as illustrated by curve I of Fig. 1, the two polymers in the blend will separate into phases that are compositionally pure in either component providing that a state of thermodynamic equilibrium has been reached. For complete miscibility, both ΔG_M must be negative and the second derivative

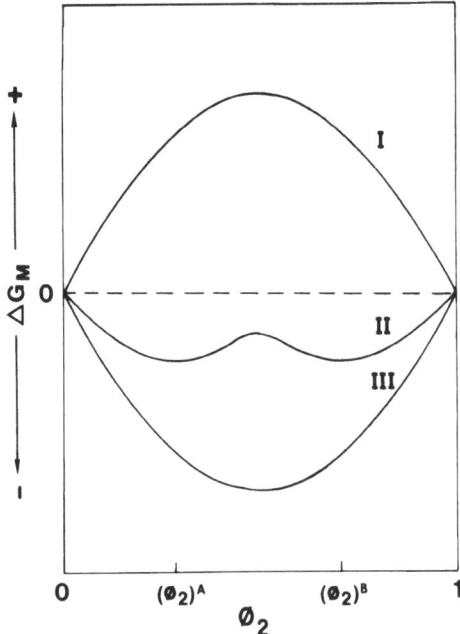

FIG. 1. Idealised isothermal plots of the Gibbs free energy of mixing (ΔG_M) versus volume fraction of component 2 (ϕ_2) in a binary polymer blend. Curves I and III represent extreme cases of thermodynamic immiscibility and miscibility, respectively. Curve II represents the intermediate case of partial miscibility whereby the blend will separate into phases whose compositions are given by the volume fraction coordinates, $(\phi_2)^A$ and $(\phi_2)^B$, corresponding to the minima in the free energy curve.

of ΔG_M with respect to the volume fraction of component 2 (ϕ_2) must be greater than zero.[8]

$$(\partial^2 \Delta G_M / \partial \phi_2^2)_{T,P} > 0 \qquad (2)$$

These conditions are satisfied by curve III of Fig. 1 but not curve II which exhibits two minima in ΔG_M and therefore the derivative criteria given by eqn (2) no longer holds. A blend whose free energy follows the composition dependence illustrated by curve II will separate at equilibrium into two phases whose compositions are given by the values of ϕ at the two minima.

The dependence of thermodynamic miscibility of a polymer pair on both composition and temperature may be represented by a liquid–liquid phase diagram like the one shown in Fig. 2. At temperature T_1, which is

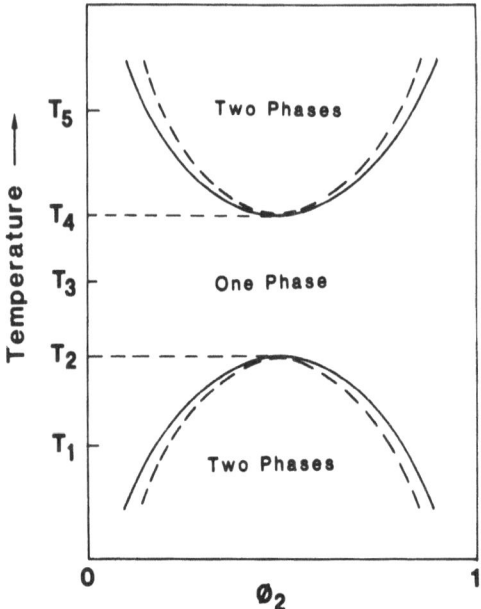

FIG. 2.   Idealised liquid–liquid phase diagram for a polymer blend showing both an upper (UCST) and lower (LCST) critical solution temperature (temperatures T_2 and T_4, respectively). Thermodynamically unstable regions are contained within the spinodal (---). The binodal (——) demarks the boundary between metastable and stable (single phase) regions.

below the upper critical solution temperature (UCST) for phase separation located at T_2, the equilibrium mixture will separate into two phases whose compositions lie on opposite sides of the binodal at T_1. The binodal separates the stable (single phase) from the metastable state while the spinodal marks the transition from the unstable to metastable region. Rules to calculate the spinodal and binodal from free energy–composition diagrams are discussed in various reviews.[2, 3, 8] At T_3, which is above the UCST but below the lower critical solution temperature (LCST) located at T_4, the blend is a thermodynamically miscible mixture at all compositions. Above the LCST, e.g. at T_5, two phases again co-exist with compositions indicated by the upper binodal.

Evidence for the existence of a LCST in several polymer blends has been cited recently.[1, 9, 10] By comparison, there is little and usually only indirect evidence for an UCST for polymer blends.[10–12] Inability to detect an UCST for a given polymer blend may result from a phase change such as crystallisation of one of the components from the melt or vitrification of the amorphous mixture at temperatures above the UCST of the blend.[13] Such a change in state would provide a diffusional barrier for the attainment of thermodynamic equilibrium and thereby prevent direct experimental observation of any low temperature binodal. A comprehensive treatment of phase behaviour in polymer blends is given by Kwei and Wang.[14]

The bulk of experimental evidence indicates that most polymer pairs are immiscible. This observation is a consequence of the very small combination entropy change (ΔS_M) which results when two high molecular weight polymers are blended. The relation of ΔS_M to the degree of polymerisation of the two blended polymers is given by the Flory–Huggins[15] lattice model for polymer solutions as applied by Scott[16] to polymer mixtures:

$$\Delta S_M = (RV/V_r)\,[(\phi_1 \ln \phi_1/x_1) + (\phi_2 \ln \phi_2/x_2)] \tag{3}$$

where R is the ideal gas constant, V_r is a reference volume, and x_1 is the degree of polymerisation of one of the blend components relative to the reference volume. Typical values for the entropic term, $T\Delta S_M$, in eqn (1) for high polymer blends may be less than 0.005 cal g^{-1}.[17] In order that ΔG_M be negative, the small negative entropic term cannot be dominated by a large positive contribution from the enthalpy of mixing, ΔH_M. In other words, polymer miscibility is limited to those blends with an exothermic or mildly endothermic heat of mixing. This implies that blend miscibility requires favourable interactions between the two polymers in the blend. Such interactions as hydrogen bonding, dipole–dipole, or charge transfer complexation that can lead to a favourable energetic state are apparently rare among high polymers and only a few dozen cases of blend miscibility have been demonstrated.

The relation of enthalpic and entropic contributions to the miscibility of polymer pairs can be viewed through the general form of the free energy expression, eqn (1), as given by Roe:[18]

$$\Delta G_M = RT[(\phi_1/\overline{V}_1) \ln \phi_1 + (\phi_2/\overline{V}_2) \ln \phi_2] + \Lambda\phi_1\phi_2 \tag{4}$$

where ΔG_M is the Gibbs free energy of mixing *per unit volume* of the mixture, $\overline{V}_1$ and $\overline{V}_2$ are the molar volumes of the polymers in the binary mixture, and Λ is a polymer–polymer interaction parameter. In this form, the term within brackets represents the combinational entropy of mixing while the second term which includes the interaction parameter contains all other contributions (both enthalpic and entropic) to the free energy of mixing. In the classical Flory–Huggins lattice model, Λ has only enthalpic significance and is related to the more familiar Flory interaction parameter, χ, through the expression

$$\Lambda = (RT/\overline{V}_2)\chi \tag{5}$$

For polymer pairs exhibiting weak (dispersive) interactions, Λ can be estimated through the use of Hildebrand[19] solubility parameters (δ) of the two polymers

$$\Lambda \approx (\delta_1 - \delta_2)^2 \tag{6}$$

The solubility parameter, which is a measure of the cohesive energy density of the material, may be determined from physical property measurements by a variety of techniques or calculated by means of chemical group parameters. The latter approach is preferred because group calculations provide a more consistent set of δ values from which comparison can be made.[3] For polymers with strong polar or hydrogen bonding character, the solubility parameters can be resolved into dispersive (van der Waals), polar, and hydrogen bonding components.[20] Compatibility may be predicted by matching polymers whose component solubility parameters are equal.[21] The application of two-dimensional solubility maps for this purpose has proved useful in a number of cases.[22]

Although useful as a first approximation, the Flory–Huggins theory has limited applicability to polymer blends. One limitation is that a LCST cannot be predicted by the theory through its failure to take into account free volume effects. A significant improvement in the theory of polymer blend miscibility has been achieved through application of the Prigogine[23] equation of state approach by Flory[24] and others.[25-28] Another statistical thermodynamic approach has been the lattice fluid theory of Sanchez and Lacombe.[29-32] In the framework of these more sophisticated approaches which employ reduced parameters of pressure, volume, and temperature, the interaction parameter of eqn (4) has both enthalpic and entropic components which depend on temperature as well as blend composition. Methods of calculating Λ from the equation of state approach have been proposed by Roe.[18] Although conceptionally satisfying, application of the equation of state theories to predict polymer–polymer miscibility has been limited by the unavailability of accurate P–V–T data such as isobaric coefficients of thermal expansion and isothermal coefficients of compressibility which are required to calculate the reduced parameters used by these theories. For the few major polymers like polymethylene and polyisobutylene for which these values are accurately known, comparison between theory and experimental results appears satisfactory at least for predictions of polymer–polymer *immiscibility.*[33]

At present, the only useful approach to assess blend *miscibility* is by means of ΔH_M measurements. Unfortunately, experimental measurement of ΔH_M for blends of high molecular weight polymers with their correspondingly high melt viscosities is not feasible.[8] In a number of cases,[34-40] ΔH_M has been estimated through measurement of heats of solution of both the unblended polymers (ΔH_i^s) and their blends in organic solvents. Application of Hess's law then allows ΔH_M to be estimated through the relation

$$\Delta H_M = a\Delta H_1^s + b\Delta H_2^s - (a + b)\Delta H_b^s \qquad (7)$$

where a and b are the moles of polymers 1 and 2, respectively. Values of ΔH_M,[39] determined by application of Hess's law for blends of poly(2,6-dimethyl-1,4-phenylene oxide) (PMMPO) and polystyrene (PS) are plotted against PS composition in Fig. 3. As expected for this compatible blend, ΔH_M is negative over the entire composition range. One difficulty with this approach, however, is that ΔH_M is typically much smaller than ΔH_i^s and therefore high precision differential microcalorimeters of the Tian–Calvet type must be employed to obtain reliable values. In addition,

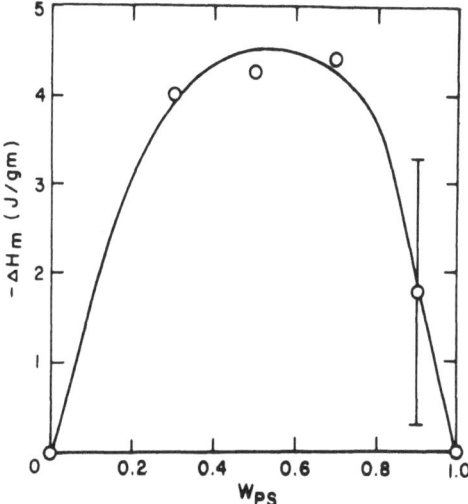

FIG. 3.   Heats of mixing (ΔH_M) as a function of weight fraction polystyrene (PS) for miscible blends of poly(2,6-dimethyl-1,4-phenylene oxide). Reproduced with permission from Ryan.[39]

large corrections for the excess enthalpy due to the initial glassy state of the polymer sample must be used to adjust ΔH_M.[36, 39, 40] A related approach has been introduced by Paul and coworkers[41, 42] who were able to infer miscibility of high polymer blends by direct measurement of ΔH_M for blends of low molecular weight, low melting compounds that are chemical analogues of the blend polymers.

1.2. Experimental Techniques
In the absence of direct thermodynamic information such as measurement of ΔH_M, blend miscibility must be inferred from the physical state of the mixture. When miscibility is inferred from indirect measurement of blend homogeneity, the term 'compatibility' appears most appropriate if it is taken to indicate homogeneity at the level of detection of the experimental technique used. Postulated differences in effective probe size afforded by different techniques have been discussed by Kaplan[43] and in a general review of solid state transitions of polymer blends by MacKnight *et al.*[44]

Of a variety of experimental criteria for blend compatibility the least conclusive is the observation of film clarity.[44] Macrophase separated blends may appear clear if the dispersed phase is smaller than the wavelength of visible light or if the two polymers comprising the blend have equal or nearly equal indices of refraction. For confirmation of a homogeneous structure, morphological examination by electron microscopy may be employed. One common difficulty with this approach is that small differences in component polymer densities or the absence of unsaturated groups that can be easily stained may make clear delineation of small phase structure difficult. Recently, small angle neutron scattering[45-47] and magic angle ^{13}C NMR techniques[48] have provided powerful techniques of probing blend homogeneity at the molecular level. Experimentally simpler but less satisfactory methods used to assess blend compatibility have included density, thermal expansion, and mechanical property measurements. As has been shown in a number of studies, all compatible blends exhibit a small but significant densification[49-57] in the solid state as a result of attractive interactions between component polymers and show enhanced tensile properties such as strength and modulus.[58-60]

Of all experimental techniques used to assess blend compatibility, measurements of the glass transition temperature (T_g) are the simplest and therefore most widely used. A single T_g intermediate between the T_gs of the unblended polymers is a necessary but sometimes inconclusive indication of blend compatibility. On the other hand, detection of two T_gs which are identical to those of the unblended polymers is a clear demonstration of immiscibility but only as it may exist at processing temperatures above the highest T_g of the blend where conditions of thermodynamic equilibrium may be approached.

Many methods exist for T_g measurement including dynamic mechanical and dielectric analysis as well as the thermal techniques of differential scanning calorimetry (DSC), differential thermal analysis (DTA), and thermal optical analysis (TOA). A discussion of the principles of thermal analysis has been given by Richardson.[61] The dynamic techniques, especially dielectric measurements, provide the advantage that frequency plane measurements may be made at various temperatures thereby providing a means of probing compatibility near a true equilibrium state provided measurements are made in the temperature region above the highest T_g of the blend.[62] Thermal analysis provides the advantage of rapid measurement and requires only small samples whose thermal history can be precisely controlled. Applications of thermal analysis to the study of blend compatibility are discussed below. For a discussion of dynamic mechanical and dielectric applications, reference is made to an earlier review.[44]

2. COMPATIBLE POLYMER BLENDS

2.1. The Glass Transition

2.1.1. Predictive Equations
The experimental dependence of compatible blend T_g on composition has frequently been modelled by relations originally applied to copolymers or polymer/diluent mixtures. Some of these have been entirely empirical or have been derived from thermodynamic or free volume arguments. The simplest of the empirical equations is the rule of mixtures[63] given as

$$T_g = W_1 T_{g,1} + W_2 T_{g,2} \tag{8}$$

where W represents the weight fraction of either polymer. Nielsen[63] has proposed a more general equation for miscible binary mixtures in the form of a modified rule of mixtures as apparently first proposed by Jenckel and Heusch[64]

$$T_g = X_1 T_{g,1} + X_2 T_{g,2} + I X_1 X_2 \tag{9}$$

where X may be volume, weight, or mole fraction of either polymer and I is an 'interaction' parameter. Other frequently used equations include the inverse rule of mixtures commonly known as the Fox[65] equation

$$(1/T_g) = (W_1/T_{g.1}) + (W_2/T_{g.2}) \tag{10}$$

and the logarithmic rule of mixtures[66]

$$\ln T_g = W_1 \ln T_{g.1} + W_2 \ln T_{g.2} \tag{11}$$

Wood[67] has proposed an equation with a single adjustable parameter, k, where

$$T_g = [T_{g.1} + (kT_{g.2} - T_{g.1})W_2]/[1 - (1 - k)W_2] \tag{12a}$$

or upon rearrangement

$$T_g = (W_1 T_{g.1} + kW_2 T_{g.2})/(W_1 + kW_2) \tag{12b}$$

The Wood parameter, k, may be related to specific parameters in several theoretically derived equations. For example, in the Gordon and Taylor[68] equation, originally derived for copolymers, k has the significance

$$k = (\Delta\alpha)_2/(\Delta\alpha)_1 \tag{13}$$

where $\Delta\alpha$ is the difference between thermal expansion coefficients above and below T_g for homopolymers corresponding to each comonomer. In the statistical mechanical theory of Gibbs and DiMarzio,[69] k is given by the expression

$$k = (n_2 M_{0.1}/n_1 M_{0.2}) \tag{14}$$

where n is the number of rotational bonds in each monomer unit and M_0 is the molecular weight of that unit.

It is interesting to note that if the empirical k parameter in the Wood equation, eqn (12), should have the value

$$k = T_{g.1}/T_{g.2} \tag{15}$$

the entire expression reduces to the simpler form of the Fox equation, eqn (10). This suggests an inherent limitation to the general applicability of the Fox equation to blends of polymers with T_gs that are not too greatly different. This conclusion is also apparent from recent theoretical considerations given later in this section.

Employing the hole theory of liquids and the isofree volume theory of the glass transition, Kanig[70] obtained an expression for the T_g of polymer/

diluent mixtures as given in the form:[71]

$$[(T_{g,1} - T_g)/(T_g - T_{g,2})\Phi_{f,2}] - (T_{g,1} - T_{g,2})/(T_g - T_{g,2}) =$$
$$C_1[\Phi_{f,1}/(T_g - T_{g,2})] + C_2 \qquad (16)$$

where

$$\Phi_{f,1} = W_1/[W_1 + (\Delta\alpha)_2 W_2/(\Delta\alpha)_1] \qquad (17)$$

and C_1 and C_2 are detailed functions of hole energies and free volume parameters.

In addition to the Kanig equation, several other approaches have employed the isofree volume theory of the glass transition to predict blend T_g. Kelley and Bueche[72] used the assumption that the specific volume of blend components are additive to derive the relation

$$T_g = [\phi_1(\Delta\alpha)_1 T_{g,1} + \phi_2(\Delta\alpha)_2 T_{g,2}]/[\phi_1(\Delta\alpha)_1 + \phi_2(\Delta\alpha)_2] \qquad (18)$$

which may be expressed in the simpler form

$$T_g = (\phi_1 T_{g,1} + k\phi_2 T_{g,2})/(\phi_1 + k\phi_2) \qquad (19)$$

where ϕ is the volume fraction of component 1 or 2 in a polymer/diluent mixture and k has the same meaning (eqn (13)) as in the Gordon–Taylor expression. In fact, except for the differences in the specification of composition, the Gordon–Taylor and Kelley–Bueche expressions are identical.

Kovacs[73, 74] has modified the Kelley–Bueche equation to allow for a nonzero excess volume of mixing, ν^E. In the form of the Kelley–Bueche expression, the Kovacs equation can be given as

$$T_g = (\phi_1 T_{g,1} + k\phi_2 T_{g,2} + k'\phi_1\phi_2)/(\phi_1 + k\phi_2) \qquad (20)$$

where the second parameter, k', is related to the fractional change in volume due to mixing $(\nu^E/\nu_{1,2})$ through the expression

$$k'\phi_1\phi_2 = \nu^E/[(\Delta\alpha)_1\nu_{1,2}] \qquad (21)$$

An expression identical in form to the Kelley–Bueche equation, eqn (19), has been derived by Somcynsky and Patterson[75] using the principle of corresponding states

$$T_g = (\phi_1 T_{g,1} + k''\phi_2 T_{g,2})/(\phi_1 + k''\phi_2) \qquad (22)$$

where the parameter, k'', is a relative measure of polymer and diluent chain flexibility.

Although several compatible polymer blends[50, 51, 76] exhibit a T_g–composition dependence which can be adequately represented by the Fox equation, eqn (10), and others[77, 78] by the Kelley–Bueche relation, eqn (19), no single equation has been used successfully in predicting the T_g of both polymer blends and mixtures of polymers with low T_g diluents. Recent thermodynamic approaches appear to offer greater promise for a unified treatment. Among these is the combined statistical mechanical and classical thermodynamic theory of Chow.[79] Using a Bragg–Williams approximation for evaluation of a configurational partition function, Chow derived an equation given in the following form

$$\ln\left(T_g/T_{g.1}\right) = \beta[(1 - \theta) \ln(1 - \theta) + \ln\theta] \qquad (23)$$

where

$$\theta = (\overline{V}_1\phi_2)/(z\overline{V}_2\phi_1) \qquad (24)$$

and

$$\beta \approx zR/(M_1\Delta C_{pp}) \qquad (25)$$

In these equations, $\overline{V}$ is the molar volume of component 1 or 2, z is a lattice coordination number, R is the ideal gas constant, M is polymer molecular weight, and ΔC_{pp} is the excess transition isobaric specific heat given as

$$\Delta C_{pp} = \Delta C_p/M_1 \qquad (26)$$

where ΔC_p is the difference in specific heat above (C_p^1) and below (C_p^g) T_g, and M is the mass of the polymer. Data for polystyrene (PS) mixed with 13 different low molecular weight diluents were found to agree well with predictions made from theory, especially at low values of θ; however, application of the Chow equation to modelling polymer blend T_g has not been made.

Gordon et al.[80] have used the configurational entropy theory of the glass transition of Gibbs and DiMarzio[81] to obtain an expression for the second order transition temperature (T_2) at which configurational entropy is zero. With the assumption that the dependence of T_g on blend composition is the same as for T_2, their equation may be written in the form of the Kelley–Bueche equation, eqn (19), as

$$T_g = (\phi_1 T_{g.1} + K\phi_2 T_{g.2})/(\phi_1 + K\phi_2) \qquad (27)$$

where K is given as

$$K = (\Delta C_{p.2}/\Delta C_{p.1}) \qquad (28)$$

Godovskii and Bessonova[82] have reported that eqn (27) afforded good fit to observed T_gs of compatible blends of PVC and nitrile–butadiene rubber (NBR). In comparison, values predicted by the Gordon–Taylor form of the Wood equation, eqn (12), and especially the inverse rule of mixtures, eqn (10), were significantly higher than those observed.

By appropriate expansion of eqn (27), Gordon et al. showed that the interaction parameter I in the generalised rule of mixtures, eqn (9), may be given in a series expansion of the form

$$I = (T_{g,2} - T_{g,1}) (1 - K) (1 + X_1 K + \ldots) \tag{29}$$

where X_1 is the mole fraction of component 1 in the mixture. Application of the Simha–Boyer[83] derived approximation that

$$T_{g,1} \Delta C_{p,1} \approx \text{constant} \tag{30}$$

to the above expansion indicates that I approaches zero when $T_{g,1} \approx T_{g,2}$. This means that in the theory of Gordon et al., the simple rule of mixtures, eqn (8), is a limiting case for the generalised rule of mixtures when the components in the blend have nearly equal T_gs (or ΔC_p values).

Couchman and Karasz[84, 85] have used a classical thermodynamic approach based on the continuity of mixture entropy at T_g to obtain an expression for the T_g of mixtures. In its most general form, it may be given as

$$\ln (T_g/T_{g,1}) = [KW_2 \ln (T_{g,2}/T_{g,1})]/[W_1 + KW_2] \tag{31}$$

where K has the same meaning (eqn (28)) as in the equation of Gordon et al. They showed that if the second and higher order terms in the series expansion of the logarithmic terms of eqn (31) are neglected, eqn (31) reduces to the form of the Gordon equation (eqn (27)):

$$T_g = (W_1 T_{g,1} + KW_2 T_{g,2})/(W_1 + KW_2) \tag{32}$$

Application of the Simha–Boyer rule, eqn (30), further reduces this relation to the Fox equation, eqn (10). If $\Delta C_{p,1} \approx \Delta C_{p,2}$, then eqn (31) reduces to the logarithmic rule of mixtures, eqn (11). Couchman[85] was able to show good agreement between T_g values predicted by eqn (31) and experimental T_gs reported for blends of PMMPO with PS and with a compatible copolymer of styrene and 4-chlorostyrene.[57] Subsequently, Leisz et al.[86] showed excellent agreement for four additional polymer blends; however, Ryan[39] has shown that experimental T_gs were substantially lower than predicted by eqn (31) in the case of one compatible blend of PMMPO and a copolymer of 2-fluorostyrene and 4-fluorostyrene (28·5%) which exhibited an atypically low $T_g \Delta C_p$ value. Fried *et al.*[87-89] have suggested that a potentially more useful form of eqn (31) is obtained by assuming only that the product $T_g \Delta C_p$ is constant as given by the Simha–Boyer rule. This simplified form may be given as

$$\ln(T_g/T_{g,1}) = W_2 \ln(T_{g,2}/T_{g,1})/[W_1(T_{g,2}/T_{g,1}) + W_2] \tag{33}$$

Fried *et al.*[87] have shown that the product $T_g \Delta C_p$ was approximately 26·1 cal g^{-1} for six polymers and four low molecular weight plasticisers in good

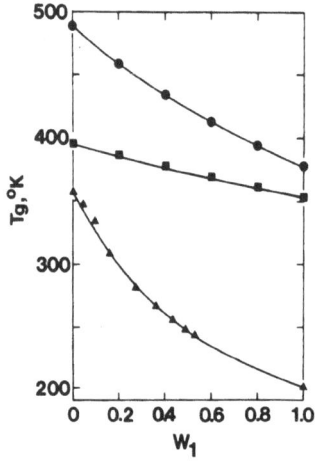

FIG. 4. DSC values of the glass transition temperature (T_g) as a function of the weight fraction of the low T_g component in compatible blends of: (●) PMMPO/ polystyrene; (■) poly(vinyl chloride)/terpolymer of α-methylstyrene, acrylonitrile, and styrene (66/31/3); and (▲) poly(vinyl chloride)/tris-(2-ethylhexyl) trimellitate plasticiser. Curves, eqn (33). Reproduced with permission from Fried.[89]

agreement with the value of 27·5 cal g^{-1} reported by Boyer[90] for 30 different polymers. The Fried equation was shown to give excellent predictions of the T_g of several polymer blends[88, 89] and mixtures of PVC with high permanence plasticisers[87] as shown by data given in Fig. 4. An advantage of eqn (33) is that values for ΔC_p are not required. This means that there is the potential of better agreement with experimentally measured blend T_gs than afforded by eqn (32) because uncertainties in the measurement of ΔC_p are eliminated.

2.1.2. Resolution of Multiple $T_g s$

There are at least two possibilities where detection of a single T_g by thermal methods can be a misleading indication of blend homogeneity.[44] The first is that two glass transitions corresponding to phases of nearly equal T_g may appear as a single, broad transition by ordinary thermal analysis. The second is that sufficiently small dispersed phases in a blend may go completely undetected by thermal methods. In either case, a micro-heterogeneous blend would be classified as compatible unless other methods are employed to confirm the absence of any thermally unresolved phases.

Glass transitions whose $T_g s$ are only 15°C apart can be resolved by careful thermal techniques as reported by Bair[91] who studied incompatible blends of PVC and a copolymer of styrene and acrylonitrile. Transitions less than 10°C apart which would ordinarily appear as a single broad glass transition may be partially resolved by annealing one phase selectively or by plotting the derivative of the C_p curve. Landi has shown the latter approach to be useful in resolving closely spaced $T_g s$ of heterogeneous copolymers of butadiene and acrylonitrile[92] for which phase separation is predicted from the Flory–Huggins theory as applied to copolymers by Scott.[93] Derivative plots also have been used by Landi to detect phase separation in blends of PVC and butadiene–acrylonitrile copolymers.[94]

Fried and Hanna[88, 89, 95, 96] have reported partial phase separation in blends of poly(2,6-dimethyl-1,4-phenylene oxide) (PMMPO) and copolymers of styrene and maleic anhydride (8%). Although these blends exhibit a single, broad glass transition by DSC, their $T_g s$ follow an unusual sigmoidal dependence on composition as shown by Fig. 5. A similarly shaped T_g–composition diagram was reported by Fried and coworkers[87] for PVC plasticised with di(2-ethylhexyl) adipate (DOA) as shown by T_g data included in Fig. 5. To provide comparison with the experimental $T_g s$ in Fig. 5, solid curves are drawn to represent $T_g s$ predicted from the Fried form (eqn (33)) of the Couchman and Karasz equation. For both the PMMPO blends and plasticised PVC, dynamic mechanical analysis was reported to indicate signs of partial phase separation evidenced by the appearance of a small shoulder on a large α (glass) relaxation peak in each spectra.

The question of the smallest phase size that thermal analysis can detect is still one of considerable controversy.[44] Recent studies by Bair and coworkers[97-100] suggest that state-of-the-art DSC techniques may be able to detect discontinuities in C_p associated with domains as small as 50–100Å. Further study of the general question of the effective probe size[43] afforded

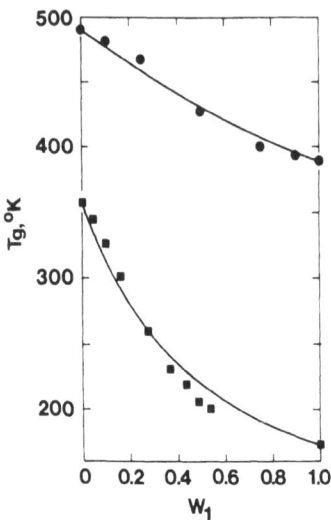

FIG. 5. DSC values of T_g as a function of the weight fraction of the low T_g component in partially compatible blends of: (●) PMMPO/poly(styrene-co-maleic anhydride); and (■) poly(vinyl chloride)/di(2-ethylhexyl) adipate. Curves, eqn (33). Reproduced with permission from Fried.[89]

by different thermal as well as dynamic methods used to determine the glass transition appears warranted.

2.2. The Crystalline Melt Temperature

Most compatible polymer blends are amorphous but several have at least one component that can be crystallised in the blend by appropriate thermal or solvent treatment. Such blends exhibit both a single compositionally dependent T_g corresponding to a mixed amorphous phase and a crystalline melting temperature (T_m), also compositionally dependent, which corresponds to a crystalline phase. Examples include blends of poly(vinylidene fluoride) (PVF$_2$) with poly(methyl methacrylate) (PMMA),[76, 101, 102] with poly(ethyl methacrylate) (PEMA),[76, 103, 104] and with poly(vinyl methyl ketone);[105] ternary blends of PVC, PMMA, and PEMA;[106] blends of PVC with poly(ε-caprolactone) (PCL)[107–112] and with various polyesters;[113] blends of bisphenol A polycarbonate with PCL[114] and with a polyester;[115] blends of PVC with a terpolymer of ethylene, vinyl acetate, and carbon monoxide;[100, 116] and blends of PMMPO with atactic and isotactic PS.[117–120] Such blends are compatible in the melt state; however, as temperature is lowered below T_m, the crystalline component can begin to crystallise.

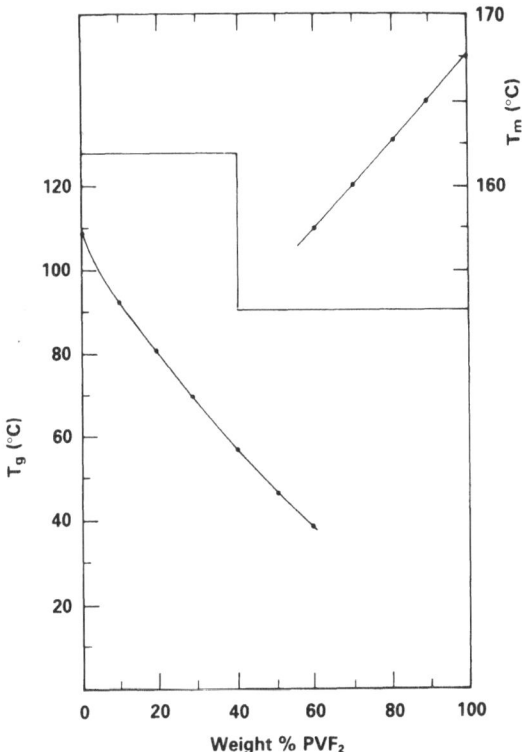

FIG. 6. Experimental values (DTA) of T_g and the crystalline melt temperature (T_m) as a function of weight percent poly(vinylidene fluoride) (PVF$_2$) in blends of PVF$_2$ and poly(methyl methacrylate) (PMMA). Reproduced with permission from Paul and Altamirano, p. 372 of reference 102b.

During the crystallisation process, the amorphous components of both polymers become incorporated within amorphous regions between pure lamellae of the crystalline component. Crystallisation of two polymers from the amorphous melt state also has been reported[105] but examples of such systems are rare.

Blending with a compatible amorphous polymer affects the crystallisation of the crystalline polymer in several ways. First, the crystallisation kinetics may be altered as a consequence of a change in the degree of undercooling during crystallisation. Figure 6 shows the effect of adding a high T_g amorphous polymer, PMMA, on the (DTA) T_g and T_m of a low T_g, crystalline polymer, PVF$_2$. Blending with PMMA increases the T_g of the mixed amorphous phase and to a smaller extent decreases the T_m of PVF$_2$.

56   J. R. FRIED

This resulting narrowing of the temperature interval, T_g–T_m, available for crystallisation, reduces the rate of crystallisation of PVF$_2$ in the blend. In part, the observed reduction in T_m may be related to the volume fraction of the amorphous polymer diluent (ϕ_1) through the equation of Nishi and Wang[101] as given in the form:[103]

$$\Delta T_m = T_m^0 - T_m \approx - T_m^0 (\bar{V}_{2u}/\Delta H_{2u}) \Lambda \phi_1^2 \qquad (34)$$

where T_m^0 is the crystalline melt temperature of the unblended crystalline polymer (component 2), T_m is its observed melt temperature in the blend, $\bar{V}_{2u}$ is the molar volume of a repeat unit of the crystalline polymer, ΔH_{2u} is the heat of fusion of the unblended crystalline polymer, and Λ is the interaction parameter as given by $(RT_m/\bar{V}_{1u})\chi$ in the form of eqn (5). For the PVF$_2$/PMMA blend, Λ has been determined to be -0.295 at 160°C.[101] Additional lowering of T_m may be attributed to morphological changes such as imperfections in crystal structure and reduction in lamellar thickness[44] although in most cases, e.g. the PVF$_2$/PMMA system, such effects may be insignificant.[13] Recently, several approaches to separate out size related effects from thermodynamic contributions to the lowering of T_m have been suggested by Stein.[121]

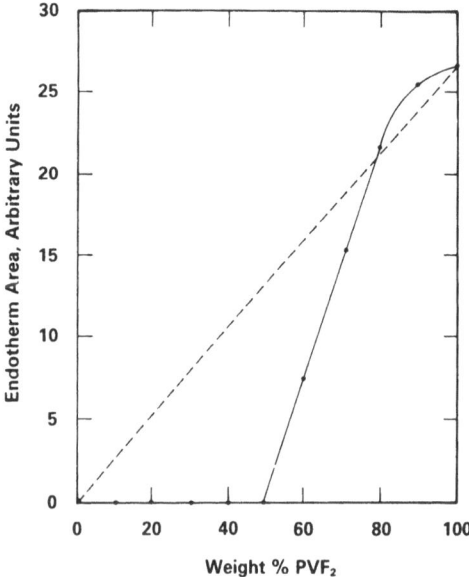

FIG. 7. Area of the normalised crystalline melting endotherm (DTA) as a function of weight percent PVF$_2$ in blends of PMMA and PVF$_2$. Reproduced with permission from Paul and Altamirano, p. 373 of reference 102b.

In the example given above, blending also limits the extent of PVF_2 crystallisation. For example, Paul and Altamirano[102] have reported that crystalline melt endotherms were not observed below 0·5 weight fraction PVF_2. At higher PVF_2 concentrations, the degree of crystallinity of PVF_2 in the blends was observed to be lower than would be expected for unblended PVF_2 subjected to identical thermal treatment. The effect of PVF_2 concentration on the area of the DTA melt endotherm (normalised for sample mass) of PVF_2 is shown in Fig. 7.

2.3. Applications of Thermal Analysis

Thermal techniques, particularly DSC, have been widely used to study diverse features of blend compatibility and to follow the development of crystallinity in crystalline–amorphous blends. For example, thermal analysis has been used by de Boer and Challa[122] to investigate the compatibility of isotactic and syndiotactic PMMA; Schneier[123] to study the effects of mixing conditions on the compatibility of PMMA and poly(vinyl acetate); Fried et al.[57, 124] to study compatibility of blends of PMMPO with copolymers of styrene and 4-chlorostyrene; Bank et al.[125] to study effects of different solvents on the compatibility of PS with poly(vinyl methyl ether) (PVME); and Frisch et al. to study interpenetrating networks (IPNs) of polyacrylates and polyurethanes[126, 127] and IPNs of PMMPO and PS.[128, 129] A complete discussion of the use of thermal techniques in the study of polymer blends is beyond the scope of this review and therefore only representative studies of approaches used by different investigators are outlined below.

2.3.1. Blend Compatibility

One of the most studied polymer blends is that of poly(2,6-dimethyl-1,4-phenylene oxide) (PMMPO) and polystyrene (PS). As previously indicated by the exothermic heat of mixing for this pair (Fig. 3) and suggested from low angle neutron scattering[46, 47] and magic angle NMR[48] measurements, blends of PMMPO and PS approach a state of thermodynamic miscibility. Among the most frequently used techniques to study the compatibility of blends of PMMPO with PS and other polymers have been thermal analysis methods, particularly DSC and thermal optical analysis (TOA). The application of other methods in the study of PMMPO/PS compatibility has been discussed in a recent review by MacKnight et al.[44]

Blends of PMMPO and PS exhibit a single, composition-dependent glass transition by DSC as has been shown by a number of investi-

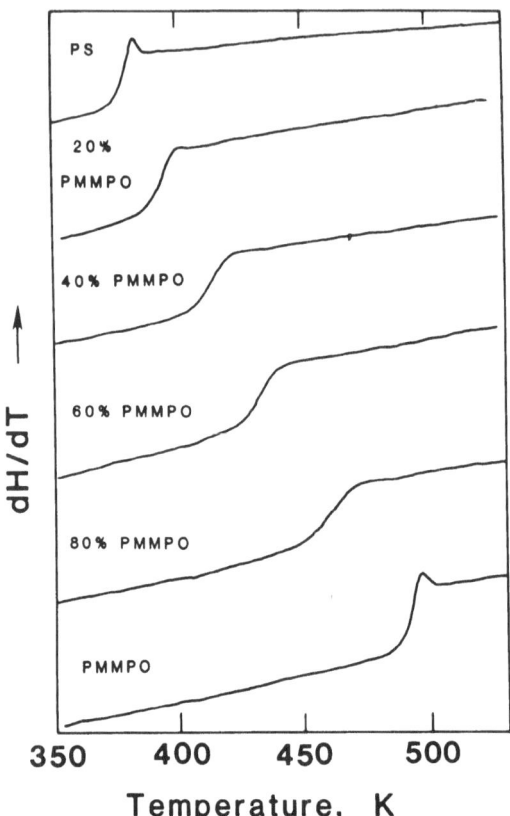

FIG. 8.   DSC thermograms of polystyrene (PS), poly(2,6-dimethyl-1,4-phenylene oxide) (PMMPO), and their blends. Thermograms are arbitrarily shifted along the ordinate to facilitate comparison. Reproduced with permission from Fried.[124]

gators.[7, 57, 124, 130–133] Representative DSC thermograms (heating rate of 20°C min^{-1}) of these blends are shown in Fig. 8. PMMPO/PS glass transition temperatures (T_gs) have been reported to follow a number of T_g–composition relations including the Fox,[131] Couchman,[85] Fried (Fig. 4), and Kelley–Bueche[133] equations. One feature evident from the DSC thermograms shown in Fig. 8 is that the blend transitions are broadened compared to those of unblended PMMPO or PS. Although such broadening appears to be common for most, if not all, known compatible blends, a firm explanation is not yet available. Broadening of the dielectric dispersion peaks of PMMPO/PS blends has also been reported[134] and attributed to localised fluctuations in blend homogeneity.[135] Possible

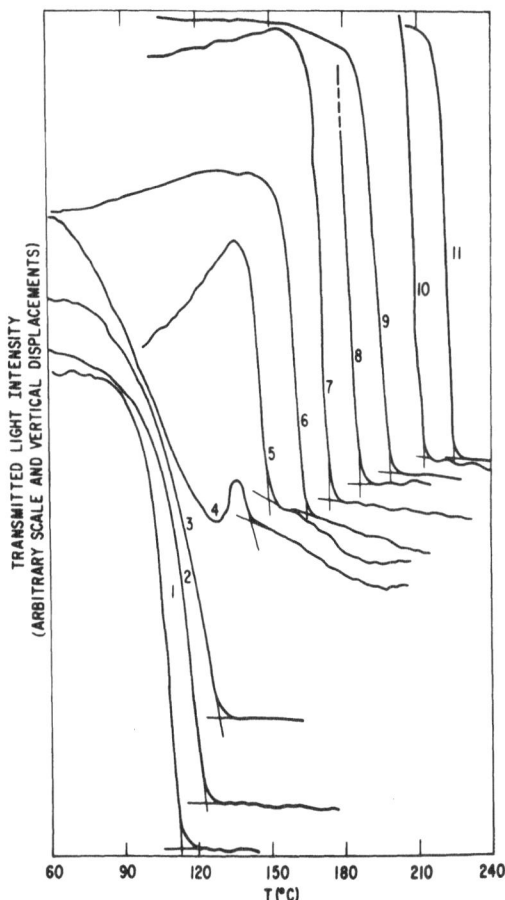

FIG. 9. Plots of transmitted light intensity (thermal optical analysis) as a function of temperature for PMMPO/PS blends. Each curve has been shifted along the ordinate (arbitrary scale) to facilitate comparison. From left to right, PS/PMMPO compositions (weight fractions) are: (1) 1/0; (2) 0·9/0·1; (3) 0·8/0·2; (4) 0·7/0·3; (5) 0·6/0·4; (6) 0·5/0·5; (7) 0·4/0·6; (8) 0·3/0·7; (9) 0·2/0·8; (10) 0·1/0·9; and (11) 0/1. Reproduced with permission from Shultz and Gendron.[131]

support of this explanation is the suggestion of molecularly sized PS domains deduced from NMR studies of the solid state.[48]

An additional feature of the thermal behaviour of these blends in common with other compatible blends is that the specific heat of each blend is a simple weighted average of the specific heats of the unblended polymers,[39, 100] i.e.,

$$(C_p^l)_b = W_1(C_p^l)_1 + W_2(C_p^l)_2 \qquad (35a)$$

$$(C_p^g)_b = W_1(C_p^g)_1 + W_2(C_p^g)_2 \qquad (35b)$$

$$(\Delta C_p)_b = W_1(\Delta C_p)_1 + W_2(\Delta C_p)_2 \qquad (35c)$$

A discussion of the corresponding specific heat behaviour of incompatible polymer blends is given in Section 3.1.

In addition to DSC measurements, thermal optical analysis (TOA) has been employed to study the behaviour of the glass transition of PMMPO/

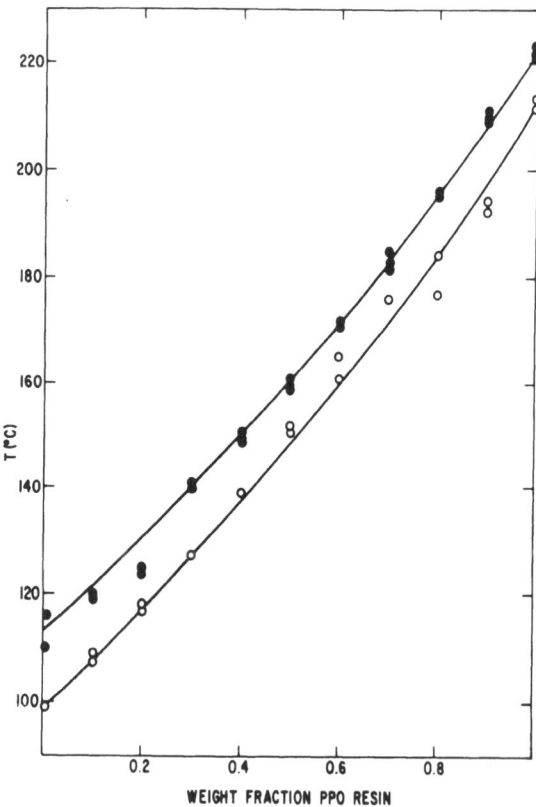

FIG. 10.   Plot of T_g (DSC) and the thermal optical transition temperature (T_{TOA}) of PMMPO/PS blends as a function of weight fraction PMMPO (PPO resin): ($\circ$), DSC T_g; ($\bullet$), T_{TOA}. Curves, eqn (10). Reproduced with permission from Shultz and Gendron.[131]

PS blends. The TOA technique as used by Shultz and coworkers[136] determines a transition temperature (T_{TOA}) as the temperature at which birefringence from closely scribed scratches on a polymer film begins to disappear. Birefringence is viewed between cross polarisers of an optical microscope which is fitted with a programmable hot stage. Plots of transmitted light intensity (arbitrary units) versus temperature for PS, PMMPO, and several PMMPO/PS blends[136] are shown in Fig. 9. In the Shultz and Gendron study, T_g was also determined by DSC (temperature at the onset of the step change in C_p at the glass transition). Values of both transition temperatures are plotted against weight fraction PMMPO in Fig. 10. Although the T_{TOA} values (10°C min^{-1}) are about 10–15°C higher than the calorimetrically determined T_gs (20°C min^{-1}), both sets of data closely follow the Fox relation, eqn (10), as shown by the curves drawn in Fig. 10. TOA has also been used by Shultz and coworkers to investigate compatibility of PMMPO with poly(2-methyl-6-benzyl-1,4-phenylene oxide);[137] with poly(4-chlorostyrene) and copolymers of styrene and 4-chlorostyrene;[132, 137, 138] with poly(2-methyl-6-phenyl-1,4-phenylene oxide);[138] and with SBS triblock copolymers.[139]

2.3.2. Interaction Parameter

As discussed in Section 1.1 the interaction parameter, Λ, is the most important factor determining polymer–polymer miscibility. Values of Λ for amorphous blends may be estimated by a number of different experimental techniques such as small angle neutron scattering,[140] inverse gas chromatography,[141-143] and vapour sorption.[144] For amorphous–crystalline polymer blends, Λ can be determined simply and rapidly by thermal measurements of the reduction of the crystalline melt temperature (T_m) through application of equations of the Nishi and Wang form, eqn (34), as discussed in Section 2.2. By this approach, values of Λ have been determined for blends of PVF$_2$ with PMMA[101] and with PEMA[102] and for blends of PMMPO and isotactic polystyrene (iPS).[120, 145]

In a systematic DTA study of the melting point depression of a variety of polyesters blended with PVC, Paul and coworkers[113] found that there was an optimum density of ester groups for achieving maximum interaction with PVC. Values of the Flory interaction parameter, χ, obtained by use of the Nishi and Wang equation were plotted as a function of the number of aliphatic carbons per ester linkage in the polyester structure. As shown by data given in Fig. 11, polyester miscibility with PVC was found to increase with decreasing ester concentration. It was suggested that a minimum in χ should be reached at some optimum carbonyl placement. Differences

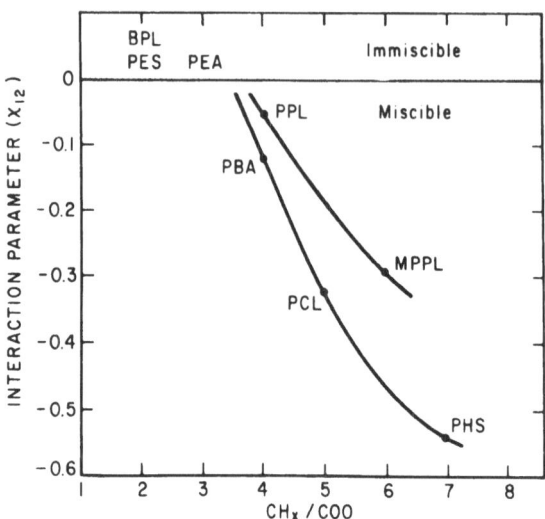

FIG. 11.   Flory interaction parameters of blends of PVC with various polyesters as a function of the ratio of the number of aliphatic carbons per carbonyl group in the polyester. Lower curve represents interaction parameters for miscible linear polyesters; upper curve for miscible polyesters with aliphatic side groups. Polyesters: BPL, poly(β-propiolactone); PES, poly(ethylene succinate); PEA, poly(ethylene adipate); PPL, poly(pivalolactone); MPPL, poly(α-methyl-α-n-propyl-β-propiolactone); PBA, poly(butylene adipate); PCL, poly(ϵ-caprolactone); and PHS, poly(hexamethylene sebacate). Reproduced with permission from Ziska *et al.*[113] by permission of the publishers, IPC Business Press Ltd.©

between χ values for polyesters with and without aliphatic side groups were attributed to carbonyl shielding effects. At high carbonyl concentrations (less than four aliphatic carbons per ester group), polyesters were found to be immiscible (i.e. $\chi > 0$) with PVC.

Values of the interaction parameter can also be determined by thermal analysis if one of the blend polymers can be made to crystallise through solvent treatment. For example, PMMPO cannot be thermally crystallised but will crystallise when exposed to solvent vapour[117] or when cast from solution. Using this approach, Shultz and McCullough[146] estimated that χ for blends of PMMPO and atactic polystyrene (aPS) is nearly zero from measurements of the melting point depression of a ternary mixture of PMMPO, aPS, and toluene. This value is in good agreement with the small negative value of ΔH_M subsequently reported for this blend by other investigators using Tian–Calvet microcalorimetry.[36, 39]

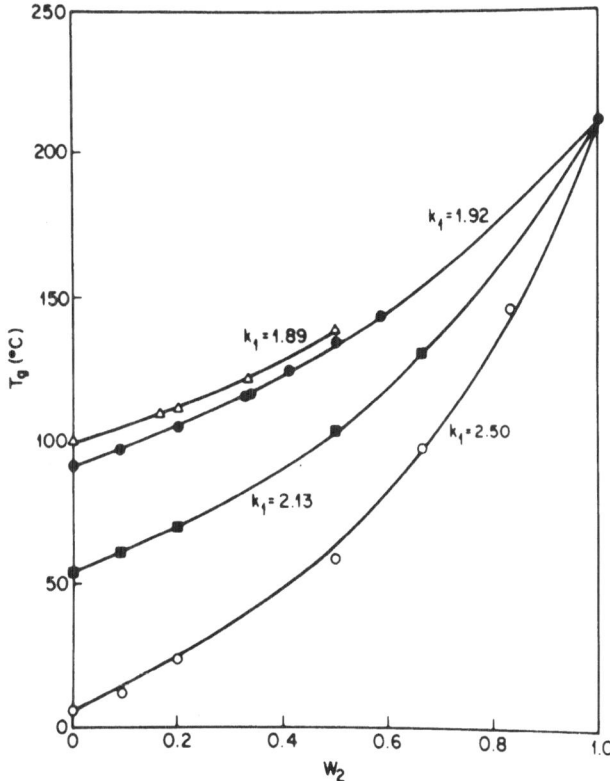

FIG. 12. Values of DSC T_g of blends of PMMPO and atactic polystyrene (aPS) as a function of both weight fraction PMMPO (abscissa) and aPS molecular weight: ($\circ$) 800 aPS; ($\blacksquare$) 2200 aPS; ($\bullet$) 10 000 aPS; and ($\triangle$) 37 000 aPS. Curves calculated from eqn (36) with values of k_1 as indicated. Reproduced with permission from Kwei and Frisch, p. 1269 of reference 147.

In a later study, Kwei and Frisch[147] directly determined Λ for PMMPO/ aPS blends in which PMMPO was crystallised by casting films from toluene solution and subsequently vacuum drying the films below T_m to remove residual solvent. The T_g and T_m of these blends were determined by DSC as the temperature at the initial rise in C_p and final temperature of the melt endotherm, respectively. Values of T_g and T_m are plotted as functions of weight fraction PMMPO and the molecular weight of aPS in the blend in Figs 12 and 13, respectively. In modelling the T_g dependence, Kwei and

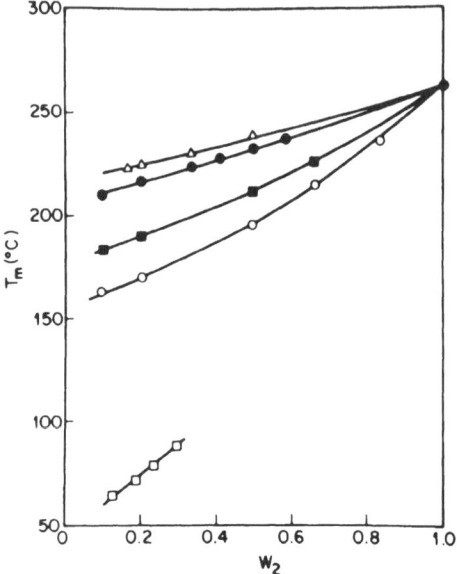

FIG. 13.   Values of DSC T_m for PMMPO/aPS blends. Identifications are the same as in Fig. 12; ($\square$) toluene. Reproduced with permission from Kwei and Frisch, p. 1269 of reference 147.

Frisch used a relation similar in form to the equations of Wood (eqn (12b)) and Gordon and Taylor:

$$T_g = (k_1 W_1 T_{g,1} + W_2 T_{g,2})/(k_1 W_1 + W_2) \tag{36}$$

where k_1 is an adjustable parameter which is a function of aPS molecular weight.

For purposes of determining χ, the equation of Nishi and Wang was extended to take into account any possible morphological contributions to the observed T_m depression. In contrast to blends of PVF_2 and PMMA where morphological effects have been found to be unimportant, small angle X-ray scattering studies of blends of PMMPO and iPS by Wenig et al.[118] have indicated a nearly linear reduction in iPS lamellar thickness with increasing PMMPO composition. The modified Nishi and Wang equation used by Kwei and Frisch was given in the form

$$\frac{\Delta H_u(T_m^0 - T_m)}{\phi_1 R T_m^0} - \frac{T_m}{M_1} - \frac{\phi_1 T_m}{2m_2} = \frac{C}{R} - b\phi_1 \tag{37}$$

where C is a morphological term, m is the polymer chain length (component 1 is the amorphous polymer, aPS), and b is a parameter used to relate the temperature dependence of χ where

$$\chi = a + (b/T) \tag{38}$$

Values of χ, determined from plots of the left side of eqn (37) vs ϕ_1, the volume fraction of aPS, were found to be a decreasing function of aPS molecular weight. A similar dependence was observed for the morphological factor, C.

2.3.3. Multicomponent Polymer Blends

Although a number of commercial blends may be complex heterogeneous mixtures containing more than two polymers for purposes of combined impact and heat distortion modification, there have been few reports of compatible polymer blends with more than two polymer components. Among useful applications of ternary systems are blends which contain AB-type block copolymers serving as compatibilising agents for otherwise incompatible binary blends of the corresponding A and B homopolymers.[148] Sometimes a single homopolymer may be compatible with two different homopolymers that by themselves are incompatible in a binary blend. An example has been given by Kwei et al.[106] who found that for certain compositions, PVF_2 can form a compatible ternary blend with PMMA and PEMA. In the ternary system, the melting point depression of PVF_2 (DSC measurements) was reported to follow the form of the unmodified Nishi and Wang equation extended to a three component system in the form

$$\Delta T_m = -T_m^0(\bar{V}_{2u}/\Delta H_{2u})[(\Lambda_{12}\phi_1 + \Lambda_{32}\phi_3)(1 - \phi_2) - \Lambda_{13}\phi_1\phi_2] \tag{39}$$

where Λ_{ij} refers to the interaction parameter for the binary pair ij. The value of Λ_{13} for the incompatible pair PMMA/PEMA was calculated to be nearly zero. Within a range of compositions, rapid quenching of the ternary blend resulted in an amorphous, compatible mixture whose T_g was found to follow a simple volume fraction average of the component T_gs as given by

$$T_g = \phi_1 T_{g,1} + \phi_2 T_{g,2} + \phi_3 T_{g,3} \tag{40}$$

It was noted that eqn (40) is a limiting form of the Kelley and Beuche expression (eqn (19)) applied to ternary blends when the thermal expansion parameter, k (eqn (13)), is equal to unity.

2.3.4. Interpenetrating Polymer Networks (IPNs)

IPNs of two incompatible, or compatible, polymers may be made by curing two cross-linkable polymers in situ. Blends formed in this manner exhibit a single, composition dependent T_g. Studies by Frisch and coworkers[126-129] have shown that the T_gs of IPNs are higher than predicted from the inverse rule of mixtures (eqn (10)) but lower than the simple rule of mixtures (eqn (8)). These investigators have suggested that the difference between the experimentally measured (DSC) T_gs and those calculated from eqn (8) can be related to the extent of interpenetration of the network through the relation

$$T_g - T_g(\text{eqn (8)}) = -[\sigma/(1+\sigma)]T_g(\text{eqn (8)}) \tag{41}$$

where σ is a parameter that is a measure of the interpenetration.[127] The maximum value of σ occurs at the maximum crosslink density, which for IPNs of PMMPO/PS was found to occur at 0·75 weight fraction PMMPO.[128, 129]

2.3.5. Polymer Adsorption

In solution, the higher molecular weight chains of many polymers such as PVC and PS will preferentially adsorb onto the surface of inorganic substrates such as silica or calcium carbonate. Adsorption can also be a competitive process between different polymers in a blend or between polymers with different tacticity. For example, Miyamoto et al.[149] have reported preferential adsorption of isotactic PMMA from a mixture with syndiotactic PMMA in chloroform solution. Thies[150] has shown that in solution with PS, PMMA can be selectively adsorbed on the surface of finely divided silica. In these studies of the competitive adsorption of two polymers in a mixture, concentrations of either polymer in both the adsorbed and nonadsorbed (solution) phase were measured by standard analytical techniques such as infrared or NMR analysis. If the two polymers form a compatible blend, then thermal measurements of T_g can be used to determine concentrations as demonstrated recently by Fried and Shih[151, 152] who studied the competitive adsorption of PMMPO and PS onto a calcium carbonate surface from a thermodynamically poor solvent, carbon tetrachloride. Concentrations of PMMPO in the solution and recovered adsorbed phase were determined by comparing the T_g (DSC) of dried, recovered fractions with an experimentally determined T_g–composition plot for PMMPO/PS blends. Results for the adsorption of a 0·50 weight fraction PMMPO blend (2 wt% in carbon tetrachloride) are illustrated in Fig. 14. The observed higher T_g of the adsorbed layer corresponds to a higher weight fraction of PMMPO adsorbed onto the calcium carbonate surface. Preferential adsorption was shown to increase as the amount of calcium carbonate available for adsorption was decreased.

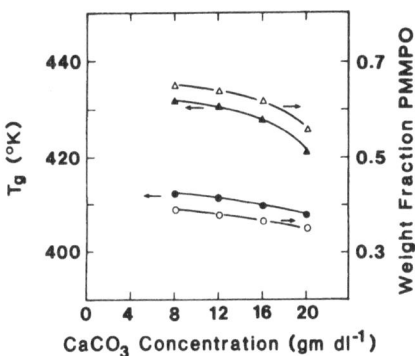

FIG. 14. Values of DSC T_g and corresponding calculated weight fraction PMMPO in the adsorbed and solution phase resulting from the adsorption of a 50/50 blend of PMMPO and PS onto calcium carbonate from carbon tetrachloride solution: (●) solution phase T_g; (▲) adsorbed phase T_g; (○) calculated weight fraction of PMMPO in solution; (△) calculated weight fraction of PMMPO in adsorbed phase. From Shih.[151]

2.3.6. Hybrid Techniques

Differential thermal analysis (DTA) has been used extensively in the study of polymer compatibility as a routine method of determining T_g and T_m. Recently, the applicability of DTA to the study of polymer blends has been extended by instrumental modification which allows for rapid measurement of the dielectric loss constant (ϵ'') over a wide range of frequencies (100–500 kHz). As reported by Akiyama et al.,[153] dielectric dissipation of the sample is detected in the form of a heat rise relative to a reference cell when an oscillating electric field is applied to the DTA sample. The dielectric loss is calculated from the DTA signal intensity (Δh), which is proportional to the heat rise, through the relation

$$\epsilon'' \propto \Delta h/(fV_0^2) \qquad (42)$$

where f is the frequency of the applied voltage (V_0). As usual in dielectric analysis, the glass transition temperature is taken to correspond to the temperature at the peak maximum (T_{max}) of a plot of ϵ'' versus temperature. In addition, activation energies for vitrification can be determined from the slope of T_{max} versus $1/f$ and contour maps can be constructed from ϵ''–f–$1/T$ plots. Features of these contour maps can be used as a sensitive tool in evaluating relative compatibility of different blends as shown by Akiyama *et al.* in the study of compatible blends of poly(vinyl nitrate), synthesised from poly(vinyl alcohol), with an ethylene vinyl acetate copolymer (EVA, 86% VA) and with poly(vinyl acetate). Glass transition temperatures of these blends, determined by standard DTA measurements and as T_{max}, were found to follow a composition dependence represented by a modified Gordon and Taylor equation.

3. INCOMPATIBLE POLYMER BLENDS

3.1. The Glass Transition
Glass transitions of incompatible polymer blends may be either totally independent of blend composition as in the extreme case of macrophase separated blends or be compositionally dependent as a consequence of either partial miscibility of both component polymers in each phase or as a result of interfacial mixing at phase boundaries for microphase separated systems. Each of these cases is considered individually below.

3.1.1. Macrophase Separation
In the extreme case of macrophase separation of two highly incompatible polymers, the T_g and ΔC_p (equal weight basis) of both glass transitions are unchanged compared to those of the corresponding unblended polymers. DSC thermograms for a representative macrophase separated blend of PMMPO and poly(4-chlorostyrene) are shown in Fig. 15. At each blend composition two transitions corresponding in temperature to those of the unblended components are clearly evident. These may be compared with the single, composition dependent transitions of the corresponding compatible blends of PMMPO and PS as shown previously in Fig. 8. Although temperature broadening of compatible blend transitions may be attributed to localised fluctuations in concentration, there is considerable controversy concerning the origin of transition broadening of the minor component transitions in macrophase separated blends. Available evidence suggests that broadening in these cases may depend on the size of the dispersed phase and probably little on the nature of any interfacial regions between the matrix and dispersed phases.

   In a recent study of block copolymers of styrene and α-methyl styrene, Gaur and Wunderlich[154] have concluded that changes in transition width

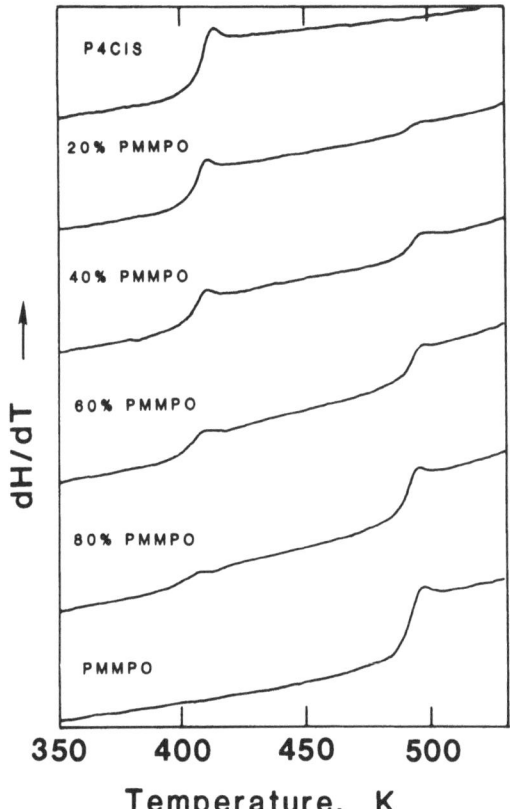

FIG. 15.   DSC thermograms of poly(4-chlorostyrene) (P4ClS), poly(2,6-dimethyl-1,4-phenylene oxide) (PMMPO), and their blends. Thermograms are arbitrarily shifted along the ordinate to facilitate comparison. Reproduced with permission from Fried.[124]

may occur when phases are less than about 0·5 μm (5000 Å) in diameter and may be independent of the nature of the matrix or interface surrounding the dispersed phase. For example, they found that the glass transitions of PS microspheres contained in a DSC sample pan (i.e. polymer–air interface) also broaden with decreasing sphere size. In these cases, broadening (ΔT_g) was evident as a lowering in temperature of the onset of the glass transition of the PS microspheres and a raising of the transition end temperature of the styrene blocks. The extent of lowering or raising in transition temperatures was found to follow the same curve when phase surface area was plotted against ΔT_g as shown in Fig. 16.

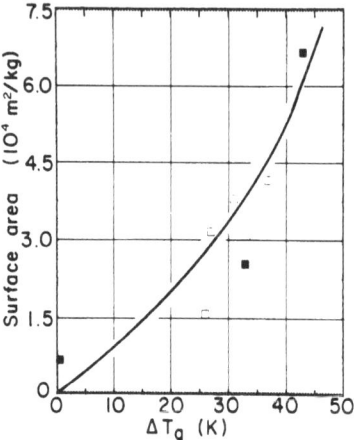

FIG. 16. Dispersed phase surface area versus extent of glass transition broadening: (■), PS microspheres; (□), lamellar microphases in block copolymers of styrene and α-methyl styrene. Reproduced with permission from Gaur and Wunderlich, p. 1623 of reference 154.

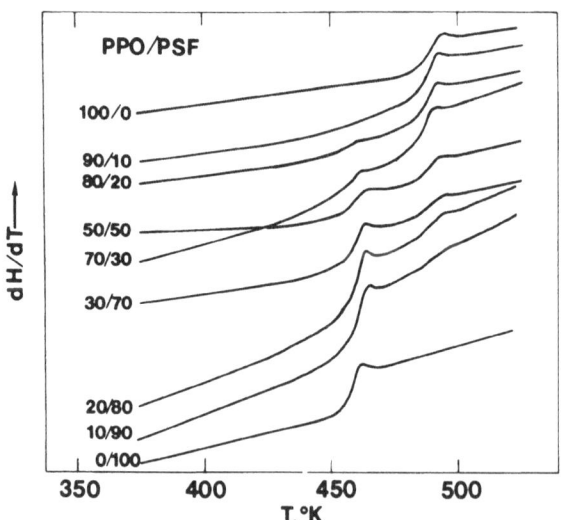

FIG. 17. DSC thermograms of poly(2,6-dimethyl-1,4-phenylene oxide) (PPO resin), bisphenol A polysulphone (PSF), and their blends. Thermograms are arbitrarily shifted along the ordinate to facilitate comparison. Reproduced with permission from Fried et al.[89]

3.1.2. Microphase Separation

As phase size decreases, minor component transitions become increasingly broadened until individual transitions of minor components may no longer be detected by routine analysis of thermal data. For example as shown in Fig. 17, the glass transition of bisphenol A polysulphone (PSF) in a 90/10 blend of PMMPO and PSF[89, 159] is marked only by a gradual increase in curvature of the C_p–temperature line near the T_g of unblended PSF (T_g = 457 K). Such extreme broadening can result in uncertainties in the assignment of T_g and an apparent low value of ΔC_p determined from routine baseline methods. Gaur and Wunderlich[154] have suggested that such indistinct transitions can be enhanced by inducing an excess enthalpy peak[155, 156] just above T_g by heating the sample at a rate faster than the one by which it was cooled. Selective annealing of one or both phases below the expected phase T_g has an equivalent effect on transition enhancement as

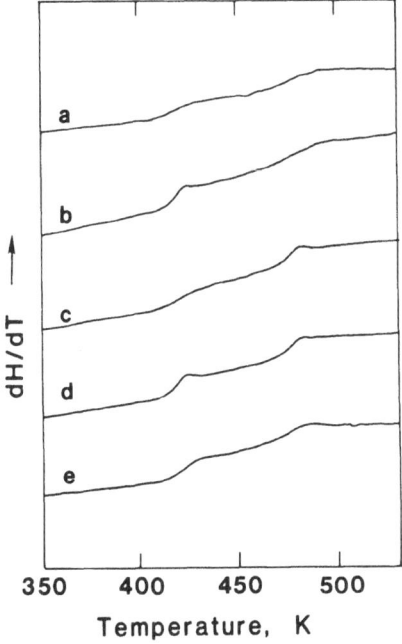

FIG. 18.   Effects of annealing on the glass transitions (DSC) of a blend of poly(2,6-dimethyl-1,4-phenylene oxide) (60 wt%) and an incompatible copolymer of styrene and 4-chlorostyrene (68·6 mole%): (a) unannealed; (b) annealed 2 hours at 393 K (below the lower T_g); (c) annealed 2 hours at 457 K (below the higher T_g); (d) annealed 2 hours at 457 K and then 2 hours at 393 K; (e) slow cooled at 0·3125°/min from 530 to 330 K (10·67 h). Reproduced with permission from Fried.[124]

shown for a 60/40 blend of PMMPO and an incompatible copolymer of styrene and 4-chlorostyrene[124] in Fig. 18.

3.1.3. Domain Size Effects

The question of whether the T_g proper of microphases may depend on domain size also has been one of recent controversy. Depression of the T_g of microphases has been reported frequently for block copolymers[158-160] and in at least one case for a polymer blend.[89, 157] Bares[158] has suggested that the T_g depression observed for the glassy microphases of thermoplastic elastomers may be attributed to interfacial mixing of the low T_g rubbery phase at the boundary of the dispersed glassy phase. Bares offered an expression for the size related T_g depression which can be given in the form

$$T_g = T_{g,i} - K_s(S/V) \qquad (43)$$

where T_g is the observed glass transition temperature of the microphase, $T_{g,i}$ is the glass transition temperature of the corresponding homopolymer of equivalent molecular weight, K_s is a constant, and S/V is the ratio of surface area to volume of the dispersed phase. More recent studies by Krause and Iskandar[159] and those by Gaur and Wunderlich[154] suggest that interfacial effects are probably insignificant for the majority of block copolymers when the block components are highly immiscible. In addition, Gaur and Wunderlich have shown that T_gs which appear depressed when determined from standard C_p-temperature plots may actually be unchanged when T_g is obtained from the more accurate but more demanding procedure of linearly extrapolating enthalpy-temperature curves above and below T_g.[61, 161, 162]

Couchman and Karasz[163, 164] have considered the role of two possible effects on the T_g of microspheres: interfacial mixing and interfacial surface tension (γ_i). From the basis of the Kelley-Bueche isofree volume model of the glass transition of polymer-diluent mixtures (eqn (19)), they obtained a general relation for the effect of interfacial mixing on the T_g of a dispersed phase in a form similar to the Bares equation and given as

$$T_g \approx T_{g,1} + k(T_{g,2} - T_{g,1})K'_s(S/V) \qquad (44)$$

where k is the Gordon-Taylor parameter (eqn (13)) and K'_s is an empirical constant. This relation includes the case not considered by Bares where the apparent T_g of the microphase may be raised due to interfacial mixing of a higher T_g matrix component (component 2). An alternative relation was obtained on the basis of the classical thermodynamic approach used by

Couchman and Karasz[84, 85] to obtain an expression for the T_g of mixtures (Section 2.1.1.). Their equation for the size related T_g depression may be given in the form

$$T_g \approx T_{g,1} + K(T_{g,2} - T_{g,1})K_s''(S/V) \qquad (45)$$

where K is given by the ratio of $\Delta C_{p,2}$ to $\Delta C_{p,1}$ (eqn (28)) and K_s'' is another constant. Couchman and Karasz further showed that the effect of internal phase pressure on T_g can be related to the interfacial tension (γ_i) through the expression

$$dT_g/dC = (\Delta\beta/\Delta\alpha)\gamma_i \qquad (46)$$

where C is the total curvature of the microphase surface while $\Delta\beta$ and $\Delta\alpha$ are differences in coefficients of compressibility and thermal expansion, respectively, as previously defined (Section 2.1.1). Since values of γ_i are typically small (1–2 dyn cm^{-1}) for polymer blends[165] and the ratio $\Delta\beta/\Delta\alpha$ is normally positive, the maximum T_g–size effect predicted by eqn (46) can be shown to be an *increase* in T_g of less than 5°C. Such small effects may not be normally noticed by routine thermal analysis.

3.1.4. Interfacial Effects

In contrast to block copolymers for which phase dimensions are controlled by block size, interfacial effects for microphase separated polymer blends cannot be routinely dismissed. Interfacial regions in blends can vary widely in size and character.[166–170] Godovskii and Bessonova[82] have considered several possible interfacial morphologies and their effects on DSC glass transitions as represented in Fig. 19. In this discussion the complicating effect of individual phase size on the observed transitions is not considered. Figure 19a represents the idealised case where the interface is conceived as a mixed phase whose composition is constant across the boundary. Separate glass transitions would be expected for each of the pure component phases and for the mixed composition interphase. In Fig. 19b, the interphase is considered to vary in composition from one phase to the other. In such a case, the intermediate interphase transition of Fig. 19a would be replaced by a continuous distribution of indistinguishably small glass transitions corresponding to the continuous distributions of compositions across the interphase. In both mixed composition interphase models, the apparent weight fraction (W) of both polymers participating in the interphase may be estimated from the relation

$$W = 1 - [(\Delta C_{p,1})^{app}/\Delta C_{p,1}] - [(\Delta C_{p,2})^{app}/\Delta C_{p,2}] \qquad (47)$$

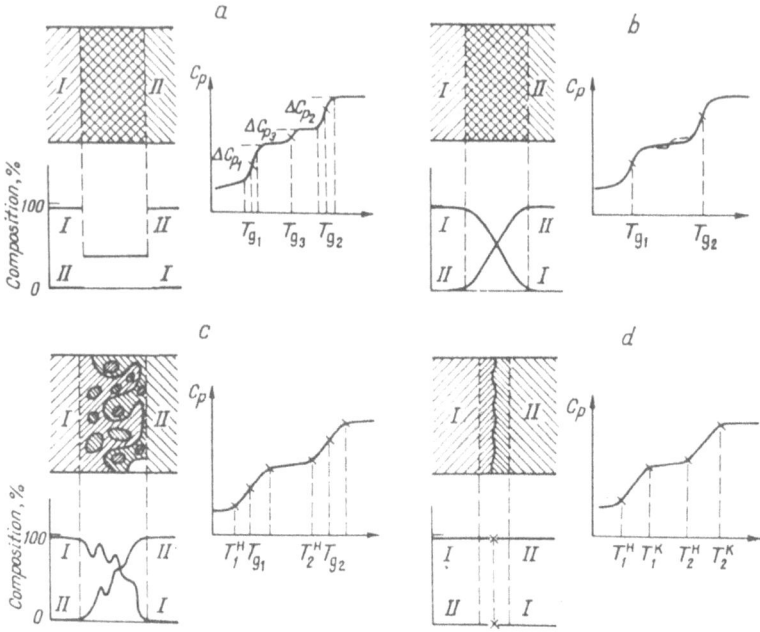

FIG. 19. Effect of interfacial morphology on the glass transition behaviour of polymer blends: (a) constant composition interface; (b) variable composition interface; (c) emulsified interface; (d) sharp interface with phase boundary layers. Reproduced with permission from Godovskii and Bessonova.[82]

where $(\Delta C_{p.1})^{\text{app}}$ is the apparent ΔC_p of component 1 in the blend. In one study of a partially miscible blend of PMMPO and a copolymer of styrene and 4-chlorostyrene, Fried *et al.*[57, 124] have suggested that as much as 40 wt% of the total blend polymers may be participating in a mixed, variable composition interface of this type.

A third case is given in Fig. 19c where the interfacial regions are pictured to contain small domains of each polymer. Lipatov[171, 172] has suggested that such a situation may occur when both polymers have equal or nearly equal surface tensions. In this case, both transitions would be expected to be broadened (shifted to lower temperatures) due to the presence of the small interfacial domains but would be undiminished due to the absence of interfacial mixing.

The final example is given by Fig. 19d which illustrates the case of two phases separated by a sharp transition region. Each phase is pictured to have a boundary layer whose density may vary from that of the bulk polymer. Density fluctuations (free volume redistribution) across the boundary regions would result in broadening of each phase transition as in the previous example.

Lipatov[173, 174] has suggested that in the absence of all interfacial effects that would alter the distribution of free volume in the blend, the Simha–Boyer relation[83, 175] should apply for each phase in the form

$$(\Delta\alpha_i^b/\phi_i) \, (T_{g,i})^b = \text{constant} \qquad (48)$$

where $\Delta\alpha_i^b$ and $(T_{g,i})^b$ are values measured for component i in the blend. Good agreement with eqn (48) has been shown in the case of incompatible blends of PS and a polyolefin.[176] Lipatov and Vilenskii[177] have observed that in the case of several incompatible blends including blends of PS and bisphenol A polycarbonate, measured values of $\Delta\alpha_i^b$ for both blend polymers were greater than would be predicted on the basis of simple additive relations. They suggested that these higher values for $\Delta\alpha_i^b$ resulted from an increase in free volume at the boundary layers in accordance with the model pictured in Fig. 19d. Unfortunately, thermal analysis of these blends was not reported.

3.2. The Crystalline Melt Temperature

As discussed in Section 2.2, the depression of the crystalline melt temperature (T_m) of a crystalline polymer blended with a compatible amorphous polymer can be explained through a dilution effect whereby the polymer–polymer interaction parameter can be estimated through equations of the Nishi and Wang form, eqn (34). Lowering of T_m may also be observed in incompatible blends as a result of decreases in lamella thickness which depends on interfacial energy between the crystalline and amorphous phases; such effects, however, may be small. For example, Paul et al.[178] have reported depressions of 3°C and 6–7°C in the T_m of PVF$_2$ when 40 wt% PVF$_2$ was blended with two incompatible polymers, PS and PCL, respectively. These compare with depressions of 9° and 20°C for compatible blends of PVF$_2$ (40%) with poly(ethyl acrylate) and poly(vinyl acetate), respectively. In another study of T_m depression, Harrison and Runt[179] observed similarly small (<5°C) T_m depressions of polyethylene single crystals imbedded in incompatible polymer matrices of widely different T_gs. In the case of the highest T_g matrix (PMMPO), they found that the high temperature (DSC) endotherm, which marks the reorganisation of the polyethylene crystals above T_m, was nearly completely suppressed due to the presence of the rigid matrix at these temperatures.

3.3. Applications of Thermal Analysis

3.3.1. Phase Separation

Phase separation in polymer blends may be classified into three categories.[10] These include crystallisation of one (crystalline–amorphous blend) or both (crystalline–crystalline blend) polymers from the amorphous melt state of a binary blend as discussed in Section 1.2. Phase separation can also occur in amorphous blends at the lower (LCST) or upper (UCST) critical solution temperature (Section 1.1) of a particular blend. Critical solution temperatures are often determined by means of cloud point measurements;[9, 180, 181] however, if a LCST lies above the highest T_g of the blend, thermal methods (particularly DSC) may be used to obtain sufficient information to approximate a phase diagram. The approach is to heat the sample in the DSC at a controlled rate to a predetermined temperature above the blend T_g and then rapidly quench below T_g. Assuming that sufficient time was allowed at the maximum temperature for a condition of thermodynamic equilibrium to exist, a binodal may be constructed by observing temperatures at which phase separation first appears. This approach has been extensively used to obtain phase information diagrams for blends of PS and poly(methyl vinyl ether) (PVME);[10, 13, 53, 180, 182, 183] PS and the polycarbonate of tetramethyl bisphenol A (TMBPA PC);[21, 184] PMMPO and halogenated polystyrenes;[185-187] and PS with halogenated PS.[185, 188]

When blends of PS and PVME are cast from toluene solution, each blend exhibits a single although broadened glass transition indicative of a compatible system as shown in Fig. 20. When heated in the DSC to 150°C and rapidly quenched, two glass transitions nearly identical to those of the unblended polymers appear as shown in Fig. 21. These results suggest the existence of a LCST near 150°C, subsequently confirmed by cloud point measurements of Nishi and Kwei.[180]

Similar observations of a LCST were made for cast blends of PS and TMBPA PC. As the blends are heated to progressively higher temperatures to above 240°C and quenched, two glass transitions begin to appear as shown in Fig. 22 for a (40/60) PS/TMBPA PC blend. For thermal treatment above 250°C, the resulting T_gs of the two phases are slightly lower than those corresponding to the unblended polymers. This observation suggests a state of partial miscibility in this temperature range. Values of experimental T_g were used to calculate compositions through the Fox equation (eqn (10)) and a phase diagram was constructed. The resulting diagram is equivalent to a binodal for PS/TMBPA PC if thermodynamic

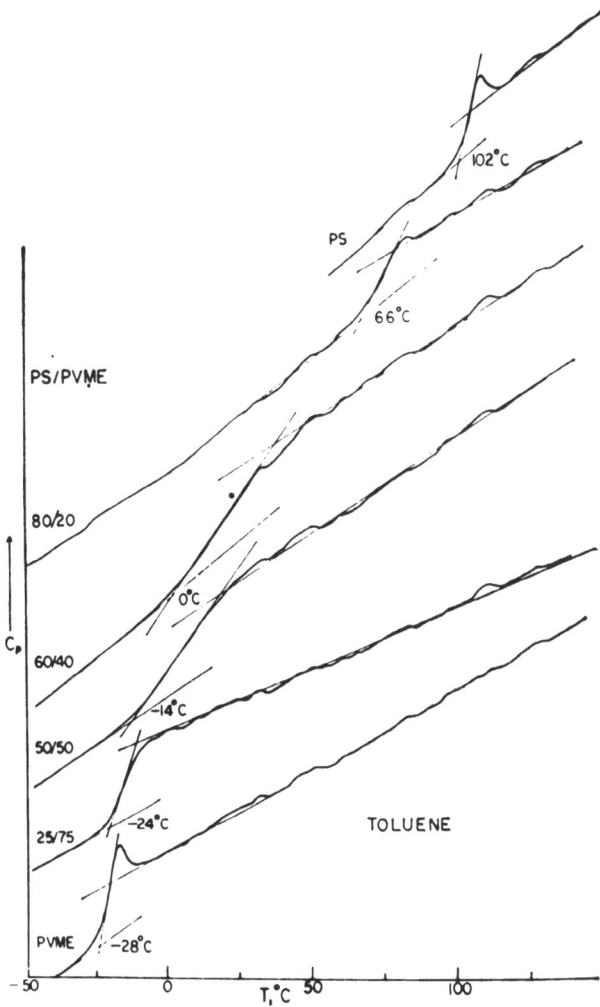

FIG. 20. DSC thermograms of toluene cast blends of polystyrene (PS) and poly(vinyl methyl ether) (PVME). Reproduced with permission from Bank *et al.* p. 44 of reference 182.

equilibrium can be assumed for the one minute thermal treatment at each T_{max}. As shown in Fig. 23, the binodal is skewed towards high concentration of TMBPA PC, the low molecular weight component in the blend.

As previously discussed (Section 2.3.1), PMMPO and PS are thermodynamically miscible over all blend compositions and exhibit single,

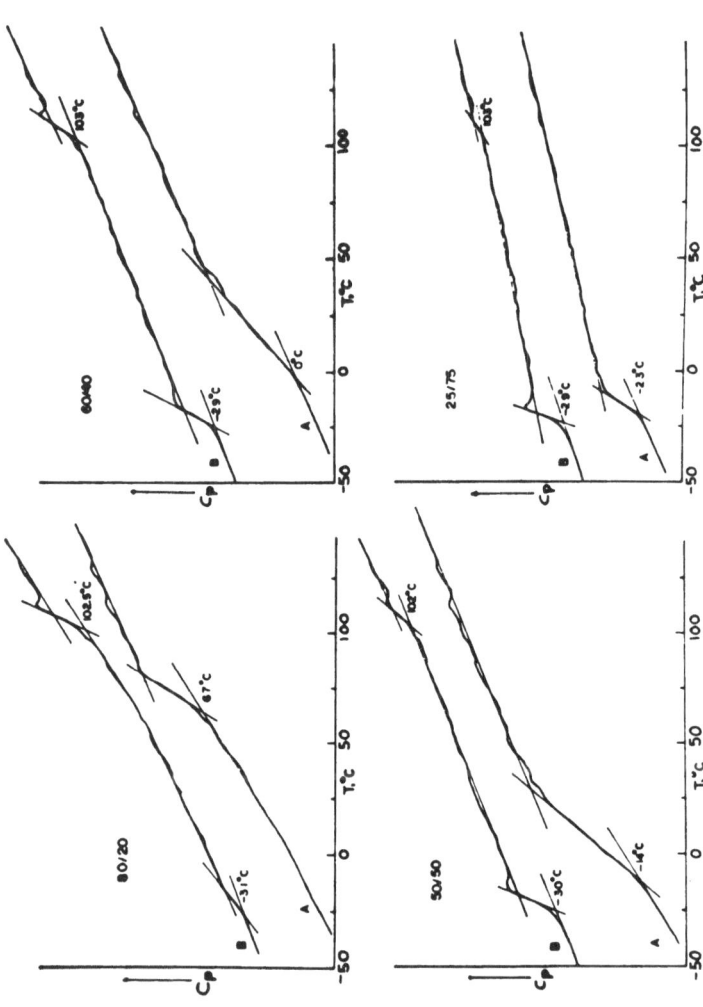

FIG. 21.   Effect of thermal treatment on the DSC thermograms of toluene cast blends of PS/PVME: (A) cast blends; (B) blends heated to 150°C and quenched. Reproduced with permission from Bank *et al.*[125]

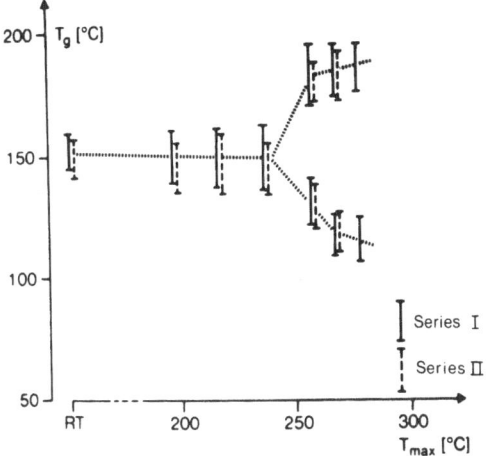

FIG. 22.   Glass transition temperatures (T_gs) of a 40/60 blend of polystyrene and tetramethyl bisphenol A polycarbonate blend heated to different temperatures (T_{max}) and quenched. Series I represents results from the initial run; Series II are repeat experiments. From Caspar and Morbitzer.[184]

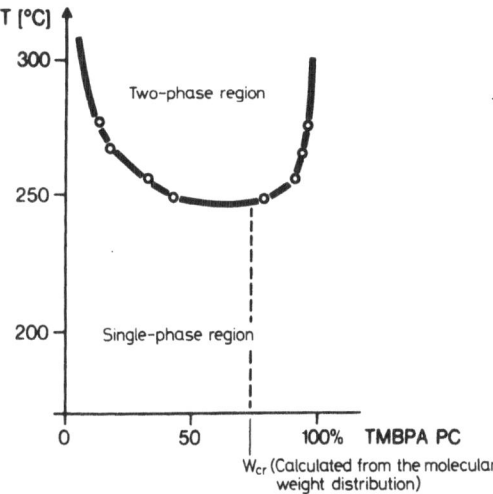

FIG. 23.   Phase diagram representing a pseudo-equilibrium binodal enclosing the unstable region for blends of polystyrene and tetramethyl bisphenol A polycarbonate. From Caspar and Morbitzer.[184]

composition dependent glass transitions as shown earlier by the DSC
thermograms in Fig. 8 and as shown for one blend composition (40 wt%
PMMPO) at the top of Fig. 24. Compatibility of the PMMPO/PS prototype
system can be conveniently altered by copolymerising styrene with various
halogenated styrenes which as homopolymers are incompatible with

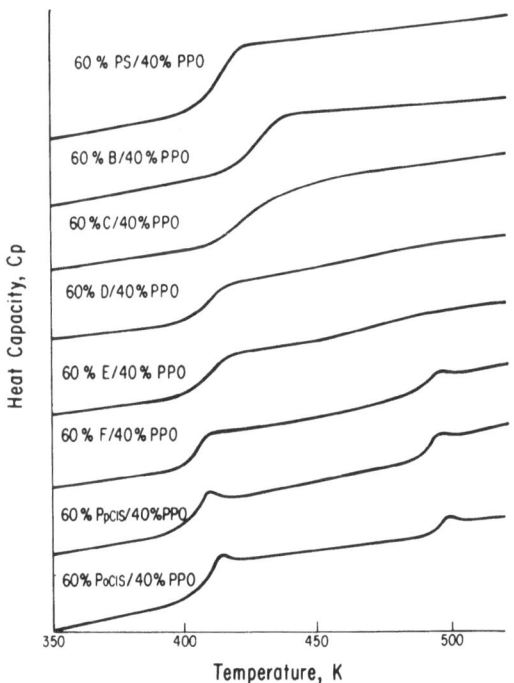

FIG. 24. DSC thermograms of blends of PMMPO (PPO) with PS, poly(2-
chlorostyrene) (PoClS), poly(4-chlorostyrene) (PpClS), and random copolymers of
styrene and 4-chlorostyrene. Copolymer compositions (mole% 4ClS); B, 58·5; C,
67·1; D, 67·8; E, 68·6; F, 75·4. Thermograms are arbitrarily shifted along the
ordinate to facilitate comparison. Reproduced with permission from Fried, p. 153
of reference 57.

PMMPO. For example, the homopolymers of poly(2-chlorostyrene) and
poly(4-chlorostyrene) are both incompatible with PMMPO as shown by
the corresponding blend thermograms (40 wt% PMMPO) at the bottom of
Fig. 24. In these examples, both transitions are sharp, undiminished, and
unchanged in T_g compared with the glass transitions of the corresponding
unblended polymers.[57, 124]

When PMMPO is blended with copolymers of styrene and 4-chloro-styrene (4-ClS), miscibility depends upon the composition of the copolymer. The transition from a miscible to immiscible blend occurs for copolymers within a narrow composition range between 67·1 and 67·8 mole% 4-ClS.[57, 124] As copolymers with decreasing 4-ClS composition are blended with PMMPO, transitions broaden, decrease in intensity, and are shifted in temperature. Transition broadening is particularly evident for PMMPO, the minor (dispersed) component, in the Copolymer D (67·8 mole% 4-ClS) and Copolymer E (68·6 mole% 4-ClS)/PMMPO blends.

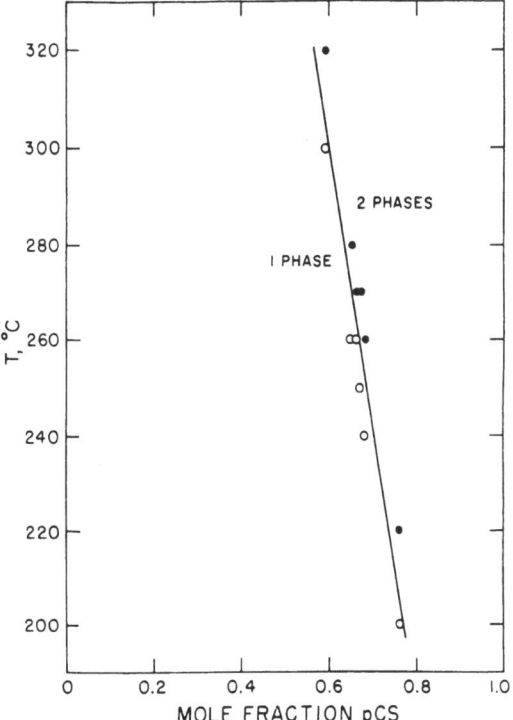

FIG. 25. Effects of thermal history and copolymer composition on phase separation of a 50/50 blend of poly(2,6-dimethyl-1,4-phenylene oxide) and copolymers of styrene and 4-chlorostyrene (pCS). The ordinate represents annealing temperature. Filled circles represent combinations of annealing temperature and copolymer composition that resulted in phase separation; open circles represent combinations that resulted in compatible blends (i.e. single T_g). The line drawn between the experimental points represents the boundary between single and two-phase systems. Reproduced with permission from Alexandrovich.[185]

These mark the transition from macrophase to microphase separated blends. The small but significant increase in the T_g of the copolymer transitions of these blends suggests partial miscibility of PMMPO in the copolymer phase—a case previously represented by curve II of Fig. 1. The glass transitions of the PMMPO phases in both of these blends also appear shifted to lower temperatures (partial miscibility with the low T_g copolymer component) although their highly broadened character make precise determination of T_g difficult.

When copolymers with 67·1 mole% or lower 4-ClS composition are blended with PMMPO, a single glass transition is observed as shown by thermograms near the top of Fig. 24 for the Copolymer B (58·5 mole% 4-ClS) and Copolymer C (67·1 mole% 4-ClS) blends. The broader transition observed for the Copolymer C/PMMPO blend may be attributed to local fluctuations in concentration within this marginally miscible blend.[135] These observations of compatibility from thermal analysis are supported by calorimetric measurements of ΔH_m which increases from $-5·31$ J g^{-1} for PMMPO/PS (50/50) blends to near zero for a blend of PMMPO and a 67·0 mole% 4-ClS copolymer.[39]

Alexandrovich[185] extended these studies by using DSC measurements to determine phase behaviour for PMMPO/copolymer blends that were heated to different temperatures and quenched. He concluded that the transition from a compatible to incompatible blend was a function not only of copolymer composition but also of the previous thermal history. Effects of both copolymer composition and thermal treatment on the phase behaviour of these PMMPO/copolymer blends are summarised by the phase diagram given in Fig. 25. LCSTs also were observed for blends of PMMPO and copolymers of 2-ClS and styrene,[185] blends of PMMPO and copolymers of 4-ClS and 2-ClS,[185–187] and blends of PS and PoClS.[185, 188]

3.3.2. *Quantitative Analysis*

Thermal analysis can be used to determine the composition of multi-component plastic resins by measurements of T_gs and intensities (ΔC_p) of individual transitions. For example, measurements of ΔC_p have been used to determine compositions of block copolymers.[189] For compatible amorphous blends, composition can be estimated from T_g–composition relations determined for that system, while for incompatible blends the weight fraction (W_i) of each component may be estimated from the ratio of observed $\Delta C_{p,i}$ to $\Delta C_{p,i}$ measured for the unblended homopolymer

$$W_i = (\Delta C_{p,i})^{obs}/\Delta C_{p,i} \qquad (49)$$

In the case of incompatible blends with one or more crystalline polymers, W_i of the crystalline component can be estimated from comparison of the observed heat of fusion (ΔQ_f) with that determined for a known composition blend with similar thermal history

$$W_i = (\Delta Q_f)^{\text{obs}}/\Delta Q_f \qquad (50)$$

Representative DSC thermograms for ABS resin are shown in Fig. 26. In the low temperature region (insert), the transition near $-90°C$ is

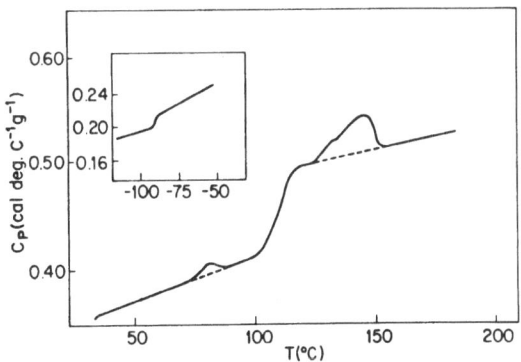

FIG. 26.   DSC thermogram of a commercial ABS resin. Insert shows the low temperature transition which corresponds to the polybutadiene component of the ABS. Reproduced with permission from Bair.[191]

identified with polybutadiene (PBD); its slightly lower T_g compared to unblended, pure PBD may be associated with the presence of about 1 wt% of styrene monomer.[190] In the high temperature range, the glass transition near 93°C is identified with a total of 76 wt% SAN copolymer while endotherms near 93 and 150°C are attributed to the melting of 0·24 wt% fatty acid residue and a moulding lubricant with respectively higher temperature transitions. Similar applications of quantitative DSC analysis have revealed detailed compositions of impact modified PVC,[191] Noryl^R modified polyphenylene oxide resins,[7] and high impact PS.[192]

Quantitative DSC analysis can also be used to determine the extent of thermal or oxidative degradation in blended resins. For example, Bair *et al.*[193] report that oxidation of PBD in commercial ABS resins can be easily followed by measurement of the decrease in ΔC_p, increase in temperature, and broadening of the glass transition of PBD.

84 J. R. FRIED

ACKNOWLEDGEMENTS

The author is grateful for many helpful suggestions received from Dr H. E. Bair of Bell Laboratories and Dr E. A. Turi of Allied Corporation who kindly reviewed the completed manuscript.

REFERENCES

1. PAUL, D. R. and BARLOW, J. W., *J. Macromol. Sci., Rev. Macromol. Chem.*, 180, **18**, 109.
2. KRAUSE, S., *J. Macromol. Sci., Rev. Macromol. Chem.*, 1972, **7**, 251.
3. KRAUSE, S., in: *Polymer Blends*, D. R. Paul and S. Newman, Eds, 1978, Academic Press, New York, Vol. 1, p. 15.
4. PAUL, D. R. and NEWMAN, S., Eds, *Polymer Blends*, 1978, Academic Press, New York.
5. OLABISI, O., ROBESON, L. M. and SHAW, M. T., *Polymer–Polymer Miscibility*, 1979, Academic Press, New York.
6. Registered trademark of the General Electric Company.
7. BAIR, H. E., *Polym. Eng. Sci.*, 1970, **10**, 247.
8. KONINGSVELD, R., KLEINTJENS, L. A. and SCHOFFELEERS, H. M., *Pure Appl. Chem.*, 1974, **39**, 1.
9. BERNSTEIN, R. E., CRUZ, C. A., PAUL, D. R. and BARLOW, J. W., *Macromolecules*, 1977, **10**, 681.
10. NISHI, T., *J. Macromol. Sci., Phys.*, 1980, **B17**, 517.
11. NISHI, T., KWEI, T. K. and WANG, T. T., *J. Appl. Phys.*, 1975, **46**, 4157.
12. WALSH, D. J., LAINGHE, S. and ZHIKUAN, C., *Polymer*, 1981, **22**, 1005.
13. KWEI, T. K., in: *Contemporary Topics in Polymer Science*, E. M. Pearce and J. R. Schaefgen, Eds, 1977, Plenum Press, New York, Vol. 2, p. 157.
14. KWEI, T. K. and WANG, T. T., *Polymer Blends*, D. R. Paul and S. Newman, Eds, 1978, Academic Press, New York, Vol. 1, p. 141.
15. FLORY, P. J., *J. Chem. Phys.*, 1944, **12**, 425.
16. SCOTT, R. L., *J. Chem. Phys.*, 1949, **17**, 1.
17. GEE, G., *Quart. Rev.*, 1947, **1**, 26.
18. ROE, R.-J., *Adv. Chem. Ser.*, 1979, **176**, 599.
19. HILDEBRAND, J. H. and SCOTT, R. L., *The Solubility of Non-Electrolytes*, 3rd ed., 1950, Reinhold, New York.
20. HANSEN, C. M., *J. Paint Technol.*, 1967, **39**, 104.
21. SHAW, M. T., *J. Appl. Polym. Sci.*, 1974, **18**, 449.
22. CABASSO, I., *Am. Chem. Soc., Div. Org. Coat. Plast. Chem., Pap.*, 1977, **37**(1), 110.
23. PRIGOGINE, I., *The Molecular Theory of Solutions*, 1957, Elsevier North-Holland Publishing Co., Amsterdam.
24. FLORY, P. J., *J. Am. Chem. Soc.*, 1965, **87**, 1933.
25. MCMASTER, L. P., *Macromolecules*, 1973, **6**, 760.
26. MCMASTER, L. P., *Adv. Chem. Ser.*, 1975, **142**, 43.

27. PATTERSON, D. and ROBARD, A., *Macromolecules*, 1978, **11**, 590.
28. LIU, D. D. and PRAUSNITZ, J. M., *Macromolecules*, 1978, **11**, 590.
29. LACOMBE, R. H. and SANCHEZ, I. C., *J. Phys. Chem.*, 1976, **80**, 2568.
30. SANCHEZ, I. C. and LACOMBE, R. H., *Macromolecules*, 1978, **11**, 1145.
31. SANCHEZ, I. C., *J. Macromol. Sci.—Phys.*, 1980, **B17**, 565.
32. SANCHEZ, I., *Polymer Blends*, D. R. Paul and S. Newman, Eds, 1978, Academic Press, New York, Vol. 1, p. 115.
33. FLORY, P. J., EICHINGER, B. E. and ORWOLL, R. A., *Macromolecules*, 1968, **1**, 287.
34. ICHIHARA, S., KOMATSU, A. and HATA, T., *Polym. J.*, 1971, **2**, 640.
35. AKIYAMA, S. and MIASA, K., *Polym. J.*, 1979, **11**, 157.
36. WEEKS, N. E., KARASZ, F. E. and MACKNIGHT, W. J., *J. Appl. Phys.*, 1977, **48**, 4068.
37. TAGAR, A. A. and BESSONOV, Y. S., *Polym. Sci. USSR*, 1975, **17**, 2741.
38. ZVEREV, M. P., POLOVIKHINA, L. A., BARASH, A. N., MAL'KOVA, L. P. and LITOVCHENKO, G. D., *Polym. Sci. USSR*, 1974, **16**, 2098.
39. RYAN, C. L., JR., *PhD Dissertation*, University of Massachusetts (Amherst), 1980.
40. RYAN, C. L., KARASZ, F. E. and MACKNIGHT, W. J., *Proc. Tenth National Thermal Analysis Society (NATAS) Conference*, October 26–29, 1980, Boston, Massachusetts, pp. 15–18.
41. CRUZ RAMOS, C. A., *PhD Dissertation*, University of Texas at Austin, 1978.
42. CRUZ RAMOS, C. A., BARLOW, J. W. and PAUL, D. R., *Macromolecules*, 1979, **12**, 726.
43. KAPLAN, D. S., *J. Appl. Polym. Sci.*, 1976, **20**, 2615.
44. MACKNIGHT, W. J., KARASZ, F. E. and FRIED, J. R., *Polymer Blends*, D. R. PAUL and S. NEWMAN, Eds, 1978, Academic Press, New York, Vol. 1, p. 185.
45. BALLARD, D. G. H., RAYNER, M. G. and SCHETTEN, J., *Polymer*, 1976, **17**, 640.
46. WIGNALL, G. D., CHILD, H. R. and LI-ARAVENA, F., *Polymer*, 1980, **21**, 131.
47. KAMBOUR, R. P., BOPP, R. C., MACONNACHIE, A. and MACKNIGHT, W. J., *Polymer*, 1980, **21**, 133.
48. STEJSKAL, E. O., SCHAEFER, J., SEFCIK, M. D. and McKAY, R. A., *Macromolecules*, 1981, **14**, 276.
49. ALEKSEYENKO, V. I., *Polym. Sci. USSR*, 1962, **3**, 367.
50. HICKMAN, J. J. and IKEDA, R. M., *J. Polym. Sci., Polym. Phys. Ed.*, 1973, **11**, 1713.
51. ZAKRZEWSKI, G. A., *Polymer*, 1973, **14**, 347.
52. JACQUES, C. H. M. and HOPFENBERG, H. B., *Polym. Eng. Sci.*, 1975, **14**, 441.
53. KWEI, T. K., NISHI, T. and ROBERTS, R. F., *Macromolecules*, 1974, **7**, 667.
54. RANBY, B. G., *J. Polym. Sci., Polym. Symp.*, 1975, **51**, 89.
55. SHUR, Y. J. and RANBY, B., *J. Appl. Polym. Sci.*, 1975, **19**, 2143.
56. YEE, A. F., *Polym. Eng. Sci.*, 1977, **17**, 213.
57. FRIED, J. R., KARASZ, F. E. and MACKNIGHT, W. J., *Macromolecules*, 1978, **11**, 150.
58. KLEINER, L. W., KARASZ, F. E. and MACKNIGHT, W. J., *Polym. Eng. Sci.*, 1979, **19**, 519.
59. FRIED, J. R., MACKNIGHT, W. J. and KARASZ, F. E., *J. Appl. Phys.*, 1979, **50**, 6052.

60. YEE, A. F. and MAXWELL, M. A., *J. Macromol. Sci.,—Phys.*, 1980, **B17**, 543.
61. RICHARDSON, M. J., in: *Developments in Polymer Characterisation*, J. V. Dawkins, Ed., 1978, Applied Science Publishers, Ltd., London, Vol. 1, p. 205.
62. WETTON, R. E., MACKNIGHT, W. J., FRIED, J. R. and KARASZ, F. E., *Macromolecules*, 1978, **11**, 158.
63. NIELSEN, L. E., *Predicting the Properties of Mixtures: Mixture Rules in Science and Engineering*, 1978, Marcel Dekker, New York.
64. JENCKEL, E. and HEUSCH, R., *Kolloid Z.*, 1953, **30**, 89.
65. FOX, T. G., *Bull. Am. Phys. Soc.*, 1956, **1**, 123.
66. POCHAN, J. M., BEATTY, C. L. and POCHAN, D. F., *Polymer*, 1979, **20**, 879.
67. WOOD, L. A., *J. Polym. Sci.*, 1958, **28**, 319.
68. GORDON, M. and TAYLOR, J. S., *J. Appl. Chem.*, 1952, **2**, 1.
69. DiMARZIO, E. A. and GIBBS, J. H., *J. Polym. Sci.*, 1959, **40**, 121.
70. KANIG, G., *Kolloid Z. Z. Polym.*, 1963, **190**, 1.
71. KRAUSE, S. and ROMAN, N., *J. Polym. Sci.*, 1965, **A3**, 1631.
72. KELLEY, F. N. and BEUCHE, F., *J. Polym. Sci.*, 1961, **50**, 549.
73. KOVACS, A. J., *Adv. Polym. Sci.*, 1963, **3**, 394.
74. BRAUN, G. and KOVACS, A. J., *Physics of Non-Crystalline Solids*, 1965, Elsevier North-Holland Publishing Co., Amsterdam.
75. SOMCYNSKY, T. and PATTERSON, D., *J. Polym. Sci.*, 1962, **62**, S151.
76. HAMMER, C. F., *Macromolecules*, 1971, **4**, 69.
77. NOLAND, J. S., HSU, N. N.-C., SAXON, R. and SCHMITT, J. M., *Adv. Chem. Ser.*, 1971, **99**, 15.
78. PREST, W. M. and PORTER, R. S., *J. Polym. Sci., Polym. Phys. Ed.*, 1972, **10**, 16.
79. CHOW, T. S., *Macromolecules*, 1980, **13**, 362.
80. GORDON, J. M., ROUSE, G. B., GIBBS, J. H. and RISEN, W. M., JR., *J. Chem. Phys.*, 1977, **66**, 4971.
81. GIBBS, J. H. and DiMARZIO, E. A., *J. Chem. Phys.*, 1958, **28**, 373.
82. GODOVSKII, Y. K. and BESSONOVA, N. P., *Polym. Sci. USSR*, 1979, **21**, 2531.
83. SIMHA, R. and BOYER, R. F., *J. Chem. Phys.*, 1962, **37**, 1003.
84. COUCHMAN, P. R. and KARASZ, F. E., *Macromolecules*, 1978, **11**, 117.
85. COUCHMAN, P. R., *Macromolecules*, 1978, **11**, 1156.
86. LEISZ, D. M., KLEINER, L. W. and GERTENBACH, P., *Thermochim. Acta*, 1980, **35**, 51.
87. FRIED, J. R., LAI, S.-Y., KLEINER, L. W. and WHEELER, M. E., *Preprints of the Third International Symposium on Polyvinylchloride*, Case Western Reserve University, Cleveland, August 10–15, 1980, pp. 186–189; *J. Appl. Polym. Sci.*, 1982, **27**, 2869.
88. FRIED, J. R., HANNA, G. A. and LAI, S.-Y., *Proc. North American Thermal Analysis Society (NATAS) Conference*, Boston, October 26–29, 1980, pp. 9–13.
89. FRIED, J. R., HANNA, G. A. and KALKANOGLU, H., in: *Polymer Compatibility and Incompatibility—Principles and Practice*, K. Solc, Ed., Harwood Academic Publishers, New York, Vol. 3 in the MMI Press Symposium Series, Midland, MI, 1982 (in press).
90. BOYER, R. F., *J. Macromol. Sci.—Phys.*, 1973, **B7**, 487.

91. BAIR, H. E., in: *Analytical Calorimetry*, R. S. Porter and J. F. Johnson, Eds, 1970, Plenum Press, New York, p. 51.
92. LANDI, V. R., *Rubber Chem. Technol.*, 1972, **42**, 222.
93. SCOTT, R. L., *J. Polym. Sci.*, 1952, **9**, 423.
94. LANDI, V. R., *Appl. Polym. Symp.*, 1974, **25**, 223.
95. FRIED, J. R. and HANNA, G. A., *Polym. Eng.* (in press).
96. HANNA, G. A., *MS Thesis*, University of Cincinnati, 1981.
97. BAIR, H. E. and WARREN, P. C, *Bull. Am. Phys. Soc.*, 1979, **24**, 288; *J. Macromol. Sci.—Phys.*, 1981, **B20**, 381.
98. BAIR, H. E. and WARREN, P. C., *Preprints of the Third International Symposium on Polyvinylchloride*, Case Western Reserve University, Cleveland, August 10–15, 1980, pp. 101–104.
99. BAIR, H. E., MATSUO, M., SALMON, W. A. and KWEI, T. K., *Macromolecules*, 1972, **5**, 114.
100. ANDERSON, E. W., BAIR, H. E., JOHNSON, G. E., KWEI, T. K., PADDEN, F. J., JR. and WILLIAMS, D., *Adv. Chem. Ser.*, 1979, **176**, 413.
101. NISHI, T. and WANG, T. T., *Macromolecules*, 1975, **8**, 909.
102. PAUL, D. R. and ALTAMIRANO, J. O., (a) *Polym. Prepr., Am. Chem. Soc., Div. Polym. Chem.*, 1974, **15**(1), 409; (b) in: *Copolymers, Polyblends, and Composites*, Norbert A. J. Platzer, Ed., Advances in Chemistry Series No. 142, American Chemical Society: Washington, D.C., 1975, pp. 371–385.
103. IMKEN, R. L., PAUL, D. R. and BARLOW, J. W., *Polym. Eng. Sci.*, 1976, **16**, 593.
104. KWEI, T. K., PATTERSON, G. D. and WANG, T. T., *Macromolecules*, 1976, **8**, 780.
105. BERNSTEIN, R. E. WAHRMUND, D. C., BARLOW, J. W. and PAUL, D. R., *Polym. Eng. Sci.*, 1978, **18**, 1220.
106. KWEI, T. K., FRISCH, H. L., RADIGAN, W. and VOGEL, S., *Macromolecules*, 1977, **10**, 157.
107. KOLESKE, J. V. and LUNDBERG, R. D., *J. Polym. Sci., Polym. Phys. Ed.*, 1969, **7**, 795.
108. ROBESON, L. M., *J. Appl. Polym. Sci.*, 1974, **17**, 3607.
109. KHAMBATTA, F. B., WARNER, F., RUSSELL, T. and STEIN, R. S., *J. Polym. Sci., Polym. Phys. Ed.*, 1976, **14**, 1391.
110. HUBBELL, D. S. and COOPER, S. L., *J. Appl. Polym. Sci.*, 1977, **21**, 3035.
111. COLEMAN, M. M. and ZARIAN, J., *J. Polym. Sci., Polym. Phys. Ed.*, 1979, **17**, 837.
112. AUBIN, M. and PRUD'HOMME, R. E., *J. Polym. Sci., Polym. Phys. Ed.*, 1981, **19**, 1245.
113. ZISKA, J. J., BARLOW, J. W. and PAUL, D. R., *Polymer*, 1981, **22**, 918.
114. CRUZ RAMOS, C. A., PAUL, D. R. and BARLOW, J. W., *J. Appl. Polym. Sci.*, 1979, **23**, 589.
115. MASI, P., PAUL, D. R. and BARLOW, J. W., in: *Rheology*, G. Astarita, G. Marrucci, and L. Nicolais, Eds, 1980, Plenum Press, New York, Vol. 3, p. 315.
116. BAIR, H. E., ANDERSON, E. W., JOHNSON, G. E. and KWEI, T. K., *Polym. Prepr., Am. Chem. Soc., Div. Polym. Chem.*, 1978, **19**(1), 143.

117. NEIRA-LEMOS, R. A., *PhD Dissertation*, University of Massachusetts (Amherst), 1974.
118. WENIG, W., KARASZ, F. E. and MacKNIGHT, W. J., *J. Appl. Phys.*, 1975, **46**, 4194.
119. HAMMEL, R., MacKNIGHT, W. J. and KARASZ, F. E., *J. Appl. Phys.*, 1975, **46**, 4199.
120. BERGHMANS, H. and OVERBERGH, N. J., *J. Polym. Sci., Polym. Phys. Ed.*, 1977, **15**, 1757.
121. STEIN, R. S., *J. Polym. Sci., Polym. Phys. Ed.*, 1981, **19**, 1281.
122. DEBOER, A. and CHALLA, G., *Polymer*, 1976, **17**, 633.
123. SCHNEIER, B., *J. Appl. Polym. Sci.*, 1974, **18**, 1999.
124. FRIED, J. R., *PhD Dissertation*, University of Massachusetts (Amherst), 1976.
125. BANK, M., LEFFINGWELL, J. and THIES, C., *J. Polym. Sci., Polym. Phys. Ed.*, 1972, **10**, 1097.
126. FRISCH, K. C., KLEMPNER, D., MIGDAL, S., FRISCH, H. L. and GHIRADELLA, H., *Polym. Eng. Sci.*, 1974, **14**, 76.
127. FRISCH, H. L., FRISCH, K. C. and KLEMPNER, D., *Polym. Eng. Sci.*, 1974, **14**, 646.
128. FRISCH, H. L., FRISCH, K. C., KLEMPNER, D. and YOON, H. K., *Am. Chem. Soc., Div. Org. Coat. Plast. Chem.*, 1979, **40**, 763.
129. FRISCH, H. L., KLEMPNER, D., YOON, H. K. and FRISCH, K. C., in: *Polymer Alloys II*, D. Klempner and K. C. Frisch, Eds, 1980, Plenum Press, New York, p. 203.
130. STOELTING, J., KARASZ, F. E. and MacKNIGHT, W. J., *Polym. Eng. Sci.*, 1970, **10**, 133.
131. SHULTZ, A. R. and GENDRON, B. M., *J. Appl. Polym. Sci.*, 1972, **16**, 461.
132. SHULTZ, A. R. and BEACH, B. M., *Macromolecules*, 1974, **14**, 902.
133. PREST, W. M., JR. and PORTER, R. S., *J. Polym. Sci., Polym. Phys. Ed.*, 1972, **10**, 1639.
134. MacKNIGHT, W. J., STOELTING, J. and KARASZ, F. E., *Adv. Chem. Ser.*, 1971, **99**, 29.
135. WETTON, R. E., MacKNIGHT, W. J., FRIED, J. R. and KARASZ, F. E., *Macromolecules*, 1974, **11**, 158.
136. SHULTZ, A. R. and GENDRON, B. M., *J. Polym. Sci., Polym. Symp.*, 1973, **43**, 29.
137. SHULTZ, A. R. and GENDRON, B. M., *Polym. Prepr., Am. Chem. Soc., Div. Polym. Chem.*, 1973, **14**(1), 571.
138. SHULTZ, A. R. and GENDRON, B. M., *J. Macromol. Sci.—Chem.*, 1974, **A8**, 175.
139. SHULTZ, A. R. and BEACH, B. M., *J. Appl. Polym. Sci.*, 1977, **21**, 2305.
140. KRUSE, W. A., KIRSTE, R. G., HAAS, J., SCHMITT, B. J. and STEIN, D. J., *Makromol. Chem.*, 1976, **177**, 1145.
141. OLABISI, O., *Macromolecules*, 1975, **8**, 316.
142. SU, C. S. and PATTERSON, D., *Macromolecules*, 1977, **10**, 708.
143. DESHPANDE, D. D., PATTERSON, D., SCHREIBER, H. P. and SU, C. S., *Macromolecules*, 1974, **7**, 530.
144. KWEI, T. K., NISHI, T. and ROBERTS, R. F., *Macromolecules*, 1974, **7**, 667.
145. RUNT, J. P., *Macromolecules*, 1981, **14**, 420.

146. SHULTZ, A. R. and McCULLOUGH, C. R., *J. Polym. Sci., Polym. Phys. Ed.*, 1972, **10**, 307.
147. KWEI, T. K. and FRISCH, H. L., *Macromolecules*, 1978, **11**, 1267.
148. ROBESON, L. M., MATZNER, M., FETTERS, L. J. and McGRATH, J. E., in: *Recent Advances in Polymer Blends, Grafts, and Blocks*, L. H. Sperling, Ed., 1974, Plenum Press, New York, p. 281.
149. MIYAMOTO, T., TOMOSHIGE, S. and INAGAKI, H., *Polym. J.*, 1974, **6**, 564.
150. THIES, C., *J. Phys. Chem.*, 1968, **70**, 3783.
151. SHIH, Y.-S., *MS Thesis*, University of Cincinnati, 1980.
152. FRIED, J. R. and SHIH, Y.-S., (unpublished data).
153. AKIYAMA, S., KOMATSU, Y. and KANEKO, R., *Polym. J.*, 1975, **7**, 172.
154. GAUR, U. and WUNDERLICH, B., *Macromolecules*, 1980, **13**, 1618.
155. PETRIE, S. E. B., *J. Macromol. Sci.—Phys.*, 1976, **B12**, 225.
156. PETRIE, S. E. B., in: *Polymeric Materials: Relationships between Structure and Mechanical Behaviour*, E. Baer and S. V. Radcliffe, Eds, 1975, American Society for Metals, Metals Park, Ohio, pp. 55–118.
157. FRIED, J. R., KALKANOGLU, H. M. and YUAN, J.-Y., *Polym. Eng. Sci.* (in press).
158. BARES, J., *Macromolecules*, 1975, **8**, 244.
159. KRAUSE, S. and ISKANDAR, M., *Adv. Chem. Ser.*, 1979, **176**, 205.
160. KRAUSE, S. and ISKANDAR, M., *Proc. Tenth North American Thermal Analysis Society (NATAS) Conference*, Boston, October 26–29, 1980, pp. 51–54.
161. FLYNN, J. H., *Thermochim. Acta*, 1974, **8**, 69.
162. RICHARDSON, M. J. and SAVILL, N. G., *Polymer*, 1975, **16**, 753.
163. COUCHMAN, P. R. and KARASZ, F. E., *J. Appl. Polym. Sci., Polym. Phys. Ed.*, 1977, **15**, 1037.
164. COUCHMAN, P. R. and KARASZ, F. E., *J. Polym. Sci., Polym. Symp.*, 1978, **63**, 271.
165. HELFAND, E. and TAGAMI, Y., *J. Chem. Phys.*, 1972, **56**, 3592.
166. VAN OENE, H. and PLUMMER, H. K., *Org. Coat. Plast. Chem.*, 1977, **37**(2), 498.
167. LIPATOV, Y. S., MOISYA, Y. G. and SEMENOVICH, G. M., *Polym. Sci. USSR*, 1977, **19**, 146.
168. LETZ, J., *J. Polym. Sci., Polym. Phys. Ed.*, 1969, **7**, 1987.
169. WOJUZKIJ, S. S., KAMENSKIJ, A. N. and FODIMANN, N. M., *Kolloid Z. Polym.*, 1967, **215**, 36.
170. LEBEDEV, Y. V., LIPATOV, Y. S. and PRIVALKO, V. P., *Polym. Sci. USSR*, 1975, **17**, 171.
171. LIPATOV, Y. S., *Polym. Sci. USSR*, 1978, **20**, 1.
172. LIPATOV, Y. S., *J. Appl. Polym. Sci.*, 1978, **22**, 1895.
173. LIPATOV, Y. S., *Polym. Sci. USSR*, 1975, **17**, 2717.
174. LIPATOV, Y. S., *J. Polym. Sci., Polym. Symp.*, 1973, **42**, 855.
175. SHARMA, S. C., MANDELKERN, L. and STEHLING, F. C., *J. Polym. Sci., Polym. Lett. Ed.*, 1972, **10**, 45.
176. BOYER, R. F. and SPENCER, R. S., *J. Appl. Phys.*, 1944, **15**, 398.
177. LIPATOV, Y. S. and VILENSKII, V. A., *Polym. Sci. USSR*, 1975, **17**, 2389.
178. PAUL, D. R., BARLOW, J. W., BERNSTEIN, R. E. and WAHRMUND, D. C., *Polym. Eng. Sci.*, 1978, **18**, 1225.

179. HARRISON, I. R. and RUNT, J., *J. Polym. Sci.*, *Polym. Phys. Ed.*, 1980, **18**, 2257.
180. NISHI, T. and KWEI, T. K., *Polymer*, 1975, **16**, 285.
181. KONINGSVELD, R., KLEINTJENS, L. A. and ONCLIN, M. H., *J. Macromol. Sci.—Phys.*, 1980, **B18**, 363.
182. BANK, M., LEFFINGWELL, J. and THIES, C., *Macromolecules*, 1971, **4**, 43.
183. NISHI, T., WANG, T. T. and KWEI, T. K., *Macromolecules*, 1975, **8**, 227.
184. CASPAR, R. and MORBITZER, L., *Angew. Makromol. Chem.*, 1977, **58/59**, 1.
185. ALEXANDROVICH, P. S., *PhD Dissertation*, University of Massachusetts (Amherst), 1978.
186. ALEXANDROVICH, P., KARASZ, F. E. and MACKNIGHT, W. J., *Polymer*, 1977, **18**, 1022.
187. KARASZ, F. E. and MACKNIGHT, W. J., in: *Contemporary Topics in Polymer Science*, E. M. Pearce and J. R. Schaefgen, Eds, 1977, Plenum Press, New York, Vol. 2, p. 143.
188. ALEXANDROVICH, P. S., KARASZ, F. E. and MACKNIGHT, W. J., *J. Macromol. Sci.—Phys.* 1980, **B17**, 501.
189. IKEDA, R. M., WALLACH, M. L. and ANGELO, R. J., in: *Block Copolymers*, S. L. Aggarwal, Ed., 1970, Plenum Press, New York.
190. REED, T. F., BAIR, H. E. and VADIMSKY, R. G., in: *Recent Advances in Polymer Blends, Grafts, and Blocks*, L. H. Sperling, Ed., 1974, Plenum Press, New York, p. 359.
191. BAIR, H. E. *Polym. Eng. Sci.*, 1974, **14**, 202.
192. BRENNAN, W. P., *Thermochim. Acta*, 1976, **17**, 285.
193. BAIR, H. E., BOYLE, D. J. and KELLEHER, P. G., *SPE Technical Papers*, 1979, **25**, 618; *Polym. Eng. Sci.*, 1980, **20**, 995.

Chapter 3

FOURIER TRANSFORM INFRARED SPECTROSCOPY OF SYNTHETIC POLYMERS

B. JASSE

Laboratoire de Physicochimie Structurale et Macromoleculaire, ESPCI, Paris, France

SUMMARY

An account is given of the applications of Fourier transform infrared (FTIR) spectroscopy to synthetic polymer systems. After a brief description of the theoretical background and advantages of FTIR spectroscopy, the developments in data processing techniques are reviewed. The computer-assisted analysis of polymer mixtures is shown to be of value in structural and conformational studies as well as in studies of polymer blends or quantitative analysis of physical and chemical changes induced in polymers. The next section considers fast-scanning FTIR spectroscopy as a tool to follow modifications occurring rapidly in polymers. Examples of dynamic changes observed during elongation and chemical reactions are reported. Finally, FTIR–photoacoustic spectroscopy is described which promises to greatly extend the usefulness of FTIR technique to 'as obtained' polymer samples.

1. INTRODUCTION

Vibrational analysis of polymers provides information on three important structural features: chemical composition, configurational and conformational structure and interatomic forces associated either with valence bonding or intermolecular interactions. Up to some ten years ago, infrared spectroscopy was mainly carried out with dispersive instruments and the limited sensitivity was an important limitation of this technique in analysing small changes in polymer systems. With the advent of Fourier transform infrared (FTIR) spectroscopy, numerous problems in the field of polymer characterisation became readily accessible on account of an increase in the signal-to-noise ratio, higher energy throughput, data processing capability and rapid scanning: an entire spectrum can be recorded in a matter of seconds. The possibility of obtaining a reasonably high quality spectrum in a matter of a few seconds opens up whole new areas of investigation which would not be possible without the aid of an interferometer system.

In the following sections, our aim is to acquaint the reader with some of the recent advances made in polymer spectroscopy due to Fourier transform infrared instrumentation. No attempt will be made to give a complete coverage of the rapidly growing FTIR spectroscopy literature and the interested reader is referred to previous reviews.[1-9]

2. THEORETICAL BACKGROUND

The basic instrument for FTIR spectroscopy is the Michelson interferometer (Fig. 1). It consists of two mirrors, one stationary and one movable, at right angles to each other, and a beamsplitter bisecting the angle between the two mirrors. The incoming radiation is partially reflected to the stationary mirror and partially transmitted to the movable mirror.

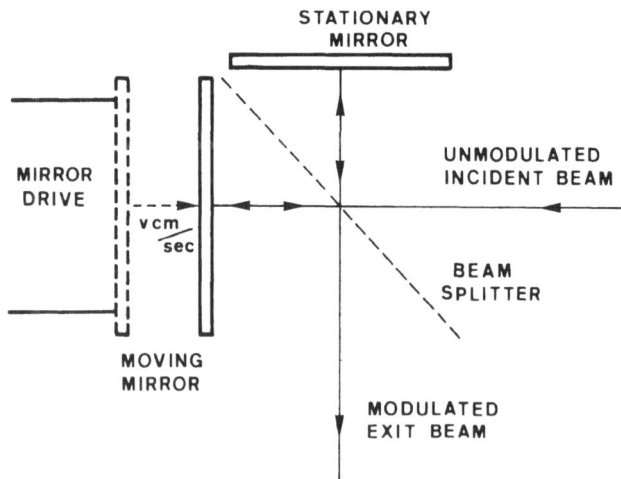

FIG. 1.   Diagram of Michelson interferometer.

After reflection from these mirrors, the two beams return to the beam-splitter where they recombine and interfere. In fact, 50% of the input beam returns along the same path and cannot be observed.

Let us consider a monochromatic source. If the fixed and movable mirrors are equidistant from the beamsplitter, no path difference exists between the two beams. They interfere constructively as well as for path differences equal to any integral number of wavelength $n\lambda$. In the case of path differences equal to $(n + 1/2)\lambda$, the two beams interfere destructively and for a monochromatic source of intensity $I(\nu)$, the intensity of the transmitted beam through the interferometer as a function of optical path difference, or retardation, x cm, is given by:

$$I(x) = 0.5I(\nu) (1 + \cos 2\pi\nu x) \tag{1}$$

In a scanning Michelson interferometer, the optical path difference x is varied by moving one mirror at a constant velocity v. The output intensity is sinusoidally modulated in dependence of time with a frequency $f(\nu) = 2v\nu$. $I(x)$ is composed of a constant component $0.5I(\nu)$ and a modulated component $0.5I(\nu) \cos 2\pi\nu x$ which is called the interferogram. In the case of a polychromatic source, the interferogram is the sum of the individual interferograms due to each wavenumber, i.e.:

$$I'(x) = 0.5 \sum_{i=1}^{n} I(\nu_i) \cos 2\pi\nu_i x \tag{2}$$

and if we consider a continuous source, the interferogram is the integral of the contributions from all wavenumbers in the spectrum:

$$I'(x) = 0.5 \int_{-\infty}^{+\infty} I(\nu) \cos 2\pi\nu x \, d\nu \tag{3}$$

In practice, the amplitude of the interferogram as observed after detection and amplification is proportional not only to the intensity of the source but also to instrument characteristics (detector response, beam-splitter efficiency, etc.). These factors remain constant for a given configuration and eqn (3) can be expressed as:

$$I'(x) = \int_{-\infty}^{+\infty} B(\nu) \cos 2\pi\nu x \, d\nu \tag{4}$$

where $B(\nu)$ represents the intensity of the source at a frequency ν cm^{-1} taking into account the instrumental characteristics. It can be seen from eqn (4) that $I'(x)$ is the cosine Fourier transform of the spectrum $B(\nu)$ which can be recovered by taking the Fourier transform of $I'(x)$:

$$B(\nu) = \int_{-\infty}^{+\infty} I'(x) \cos 2\pi\nu x \, dx \tag{5}$$

In practice, data acquisition involves signal-averaging of interferograms and requires that the signals are added coherently. For this purpose the interferogram of a monochromatic source (He–Ne laser) is measured along with the main interferogram. The different interferograms can be digitised at exactly the same position during each scan by sampling at each zero value of the sinusoidal reference interferogram arising from the He–Ne laser. It is, however, necessary for the first data point to be sampled at an identical retardation for every scan. This is achieved using a third

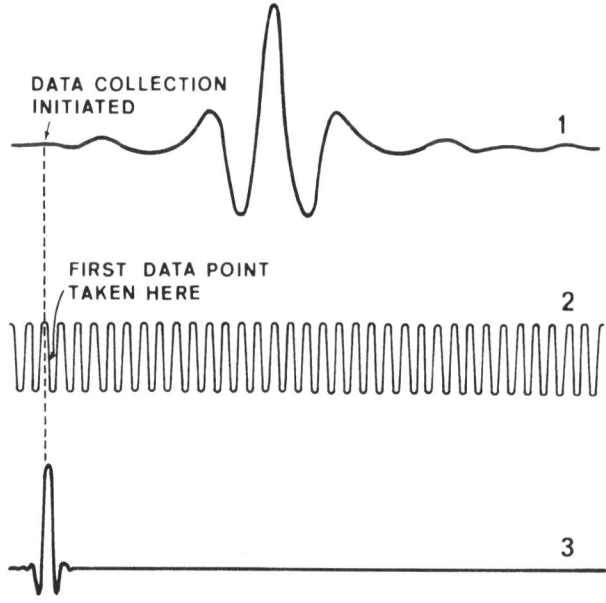

FIG. 2. Signals from an interferometer: (1) infrared interferogram; (2) laser interferogram; (3) white light interferogram.[10]

interferogram arising from a white light source. The very sharp interferogram produced by this source gives a reproducible marker at the same retardation, and whenever this interferogram exceeds a certain threshold voltage, data collection begins at the next zero crossing of the laser reference interferogram. Figure 2 illustrates such a process. A schematic description of the optical lay-out of an FTIR spectrometer is shown in Fig. 3.

The use of a computer is essential for FTIR spectroscopy. The

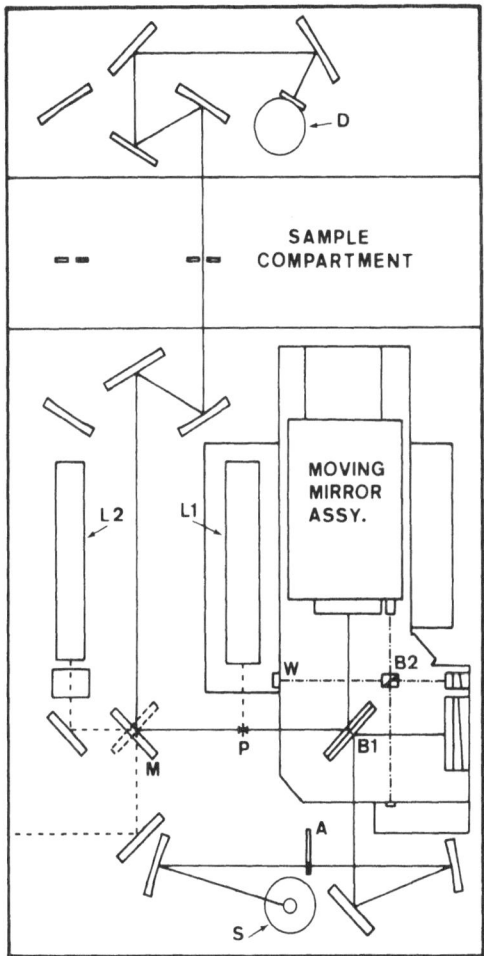

FIG. 3. Optical layout of a commercial FTIR spectrometer.[11] S–source; A–aperture; B 1–beamsplitter and compensator; B 2–white light beamsplitter; W–white light source; P–centreline laser prism; M–4-position computer-controlled flat mirror; L 1–centreline laser; L 2–alignment laser; D–infrared detector.

publication in 1965 of the Cooley–Tukey algorithm[12] allowing a fast Fourier transform computation, and the development of mini-computers over the past decade, have been essential factors in the commercialisation of FTIR spectrometers working over the entire infrared range. A detailed treatment of theory and instrumentation has been published by Griffiths.[10]

2.1. Advantages of FTIR Spectroscopy

In contrast to usual dispersive infrared spectrometers for which the radiation is divided into frequency elements by the use of a grating monochromator and slit system, all the spectral information is included in the interferogram obtained from a single scan of the movable mirror of the interferometer. The interferometer contains no slits and the amount of energy falling on the detector is greatly enhanced compared to dispersive systems (Jacquinot's advantage). In theory, Jacquinot's advantage, which depends on the resolution, may be as much as 80 to 200 times greater than that of a dispersive instrument. This advantage is particularly interesting for the study of optically dense materials (such as carbon black filled, highly coloured or absorbing polymers).

The multiplex or Fellgett's advantage arises from the fact that all spectral frequencies are measured at the same time in one scan of the interferometer. It results in an important increase of signal-to-noise ratio as compared with a dispersive instrument for identical measurement times.

Besides the previous advantages, the availability of computers is certainly the main reason for the tremendous development of FTIR spectroscopy in the polymer field. The user has the possibility of signal averaging to increase signal-to-noise ratio, and data processing for analysis of complex systems. In addition, the rapid scanning capability of the interferometer has allowed the recent development of analysis of chemical or physical structural changes in polymers as a function of time over the entire mid-infrared frequency range.

In the following sections, our aim is to acquaint the reader with some of the recent advances made in polymer characterisation due to the advantages brought about by FTIR instrumentation.

3. DATA PROCESSING TECHNIQUES

Polymer systems are often complex. For example, crystalline and amorphous regions coexist in semi-crystalline polymers. In addition different conformational structures may be observed in these regions. Any physical or chemical treatment of a polymer will induce structural changes, the knowledge of which is essential for a better understanding of polymer properties.

Quantitative analysis of such complex systems by data processing of digitised spectra has been recently developed by Koenig and coworkers.[13] The logic pattern to be followed in developing the spectral methods is

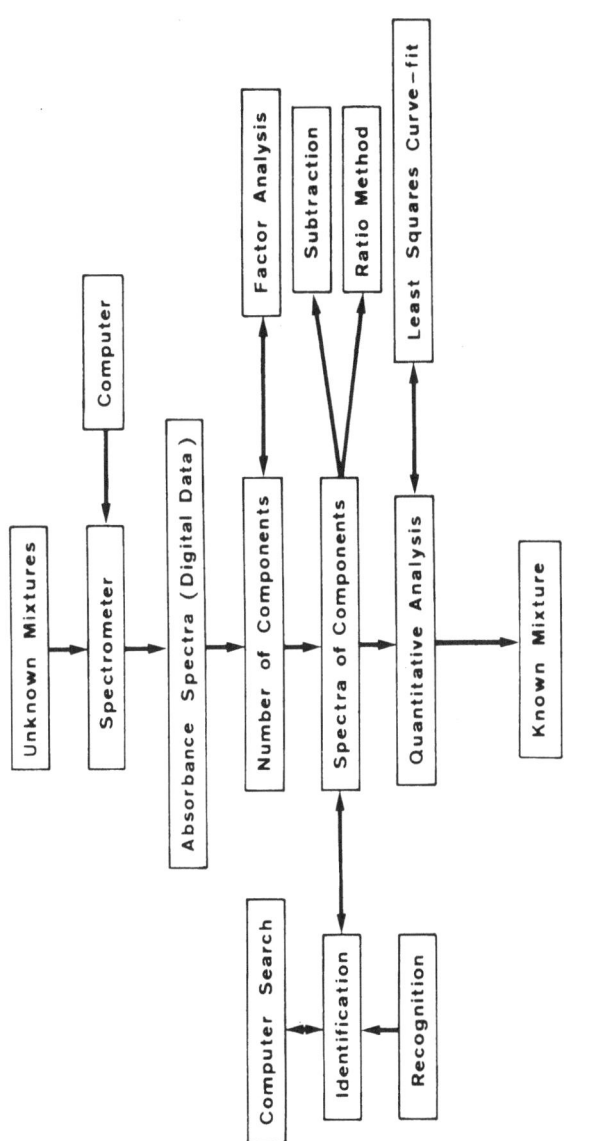

FIG. 4. Computer-assisted analysis of polymers.[13]

shown in Fig. 4. The FTIR spectrum of the mixture is obtained and stored as a digital absorbance spectrum.

3.1. Factor Analysis

The first problem to be solved concerns the number of spectroscopically differentiable components in the mixture. A method known as factor group analysis[14] has been developed for this purpose. Such a method generally applies to systems of data in which the ith measured property S_{ij} can be expressed as a linear combination of the properties and concentrations of the n individual species present:[15]

$$S_{ij} = \sum_{k=1}^{n} S_{ik} \cdot C_{kj} \qquad (6)$$

where S_{ij} is the ith measured property for the jth mixture. S_{ik} is the contribution of the kth component to the ith property and C_{kj} is the concentration of the kth component in the jth mixture. The sum is over all the components. For an infrared spectrum of a mixture, Beer's law, which holds for the relationship between absorbance and concentration has the same form:

$$A_{\nu j} = \sum_{k=1}^{n} \epsilon_{\nu k} C_{kj} \qquad (7)$$

where $A_{\nu j}$ is the absorbance for the jth mixture at wavenumber ν, $\epsilon_{\nu k}$ is the extinction coefficient of species k at the νth wavenumber, C_{kj} is the concentration of the kth species in the jth mixture and the sum is over the n absorbing species.

In matrix notation, we obtain:

$$A = \epsilon C \qquad (8)$$

The dimensions of A are $m \times n$ where m represents a given mixture spectrum and n the absorbance for each spectral element (in cm^{-1}) for the spectrum m. The problem is to obtain the rank of the A matrix, i.e. the number of linearly independent spectra in the mixture. This aim is realised by first multiplying A by A^T (A transpose) to obtain a matrix C called the covariance matrix. The dimensions of C are $m \times m$ and its rank is equivalent to the rank of A. Eigenvalues and eigenvector for C can be calculated by diagonalisation. The rank of C is equal to the number of non-zero eigenvalues of C which is also the number of components in the mixture. Experimentally some problems occur in selecting non-zero eigenvalues because of small positive eigenvalues arising from spectral noise.

The feasibility of the method was demonstrated by an analysis of *o*-, *m*-, and *p*-xylene solutions. Factor analysis has since been applied successfully to different polymer systems.[14] The method is particularly attractive in the area of conformational analysis of polymers on account of the possibility of determining the number of independent conformers. This is illustrated by the study of annealed amorphous poly(ethylene terephthalate) (PET)

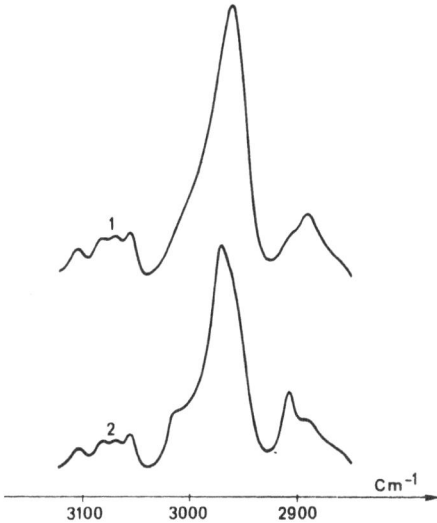

FIG. 5.   PET C—H stretching region: (1) unannealed film; (2) film annealed at 160°C for two hours.[13]

films.[13] Figure 5 shows the spectrum of the amorphous polymer and the spectrum of the same sample after annealing two hours at 160°C. A shoulder at 2909 cm^{-1} clearly increases as does an absorption band at 3015 cm^{-1}. During annealing, one can expect an increase in the amount of the *trans* conformer of the methylene parts of the polymer chain related to changes in the amorphous state and/or crystallisation. Figure 6 shows the logarithm of eigenvalues versus components obtained by applying factor analysis to a set of eight spectra of amorphous PET annealed at 200°C for 3 hours. Two eigenvalues are clearly different from the error eigenvalues arising from the signal-to-noise ratio of the spectra. As demonstrated by

Koenig *et al.*[14] the first two eigenvalues will be taken as non-zero and the others assumed to be zero due to experimental error. The two components were assigned to the *trans* and *gauche* conformational isomers of PET chains. Some crystallisation occurs under the annealing conditions used and the *trans* component must be present in both the amorphous and crystalline phases.

A similar treatment has been successfully applied to polystyrene–poly(2,6-dimethyl 1,4-phenylene oxide) (PS–PDMPO) compatible blends.[14]

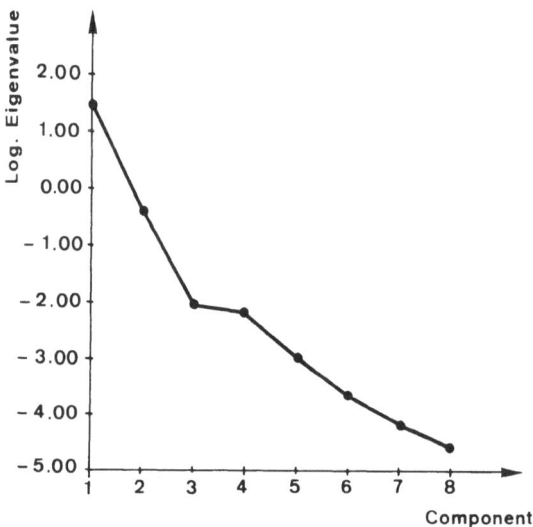

FIG. 6.   Factor analysis of annealed PET films for 3120–2850 cm^{-1} region.[13]

Seven samples were prepared with different proportions of PS and PDMPO. Factor analysis in the C—H stretching region (3200–2100 cm^{-1}) revealed that the spectra were a linear combination of the two spectra of PS and PDMPO but the same treatment in the 1800–1100 cm^{-1} region gave three linearly independent components. This result is significant of a conformational transition of PDMPO that is a function of its concentration in the blend.

A more complex system is represented by an epoxy resin crosslinked with an anhydride agent. A two component system was observed from factor analysis in the C—H stretching region (3200–2800 cm^{-1}) and in the C≡C stretching region. However if a tertiary amine is added as a catalyst, the spectrum is no longer a linear combination of the initial reactant spectra. Three components are present as shown in Fig. 7. Two of the three spectral components were assigned to pure epoxy and pure anhydride agent. The third one was thought to be characteristic of the spectral changes accompanying the reaction.

A much more complex system is represented by atactic poly(vinyl chloride) (PVC). This polymer is semi-crystalline and it is expected to be composed of two phases, crystalline and amorphous. Factor analysis[14] of a set of 10 PVC spectra in the 1470 to 1400 cm^{-1} region [$\delta(CH_2)$ mode] revealed the presence of only two components. It was concluded that the

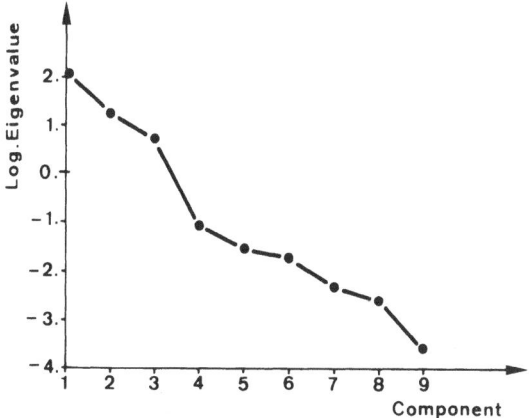

FIG. 7.   Factor analysis of a crosslinked epoxy resin for 2000–1400 cm^{-1} region.[14]

trans syndiotactic structure, regardless of its length or inclusion in the amorphous or crystalline phase, gives rise to only one independent spectral component (absorption band at 1427 cm^{-1}) while the amorphous phase is characterised by an absorption band at 1435 cm^{-1}. Therefore the $\delta(CH_2)$ mode of the amorphous phase seems to be conformationally insensitive. In contrast, factor analysis of the conformationally sensitive $\nu(C-Cl)$ region (700–550 cm^{-1}) indicates the presence of many independent components arising from highly coupled conformational modes. In conclusion, although the determination of the number of significant components is somewhat subjective, factor analysis represents an interesting method of estimating the number of components or conformational isomers contributing to the spectra.

3.2. The Ratio Method

Once the number of components in a polymer spectrum is known, it is of great interest to be able to obtain the pure spectra of these components. For this purpose, Koenig et al. developed the so-called ratio method[16, 17] for analysis of binary mixtures.

The absorbance spectrum of such a mixture can be written:

$$A = l(\epsilon_1 C_1 + \epsilon_2 C_2) \tag{9}$$

where l is the pathlength, ϵ_1 and ϵ_2 the extinction coefficients of components 1 and 2 respectively, and C_1 and C_2 the concentrations defined as mole fractions such that:

$$C_1 + C_2 = 1 \tag{10}$$

For another mixture of different concentration, the absorbance spectrum is:

$$A' = l(\epsilon_1 C_1' + \epsilon_2 C_2') \tag{11}$$

The ratio of the two spectra A and A' is given by:

$$\frac{A'}{A} = \frac{\epsilon_1 C_1' + \epsilon_2 C_2'}{\epsilon_1 C_1 + \epsilon_2 C_2} \tag{12}$$

In a frequency region where the spectral contribution of component 2 is absent, at a wavenumber ν_1 for example: $\epsilon_1 \neq 0$, $\epsilon_2 = 0$ and

$$\left(\frac{A'}{A}\right)_{\nu_1} = \left(\frac{\epsilon_1 C_1'}{\epsilon_1 C_1}\right)_{\nu_1} = \frac{C_1'}{C_1} = R_{\nu_1} \tag{13}$$

In the same way, for component 2, at a wavenumber ν_2 : $\epsilon_1 = 0$, $\epsilon_2 \neq 0$ and

$$\left(\frac{A'}{A}\right)_{\nu_2} = \left(\frac{\epsilon_2 C_2'}{\epsilon_2 C_2}\right)_{\nu_2} = \frac{C_2'}{C_2} = R_{\nu_2} \tag{14}$$

According to (10) $C_1 + C_2 = C_1' + C_2' = 1$, we obtain:

$$C_1 = \frac{1 - R_{\nu_2}}{R_{\nu_1} - R_{\nu_2}} \qquad C_1' = R_{\nu_1} C_1 \tag{15}$$

$$C_2 = \frac{R_{\nu_1} - 1}{R_{\nu_1} - R_{\nu_2}} \qquad C_2' = R_{\nu_2} C_2 \tag{16}$$

Since A and A' are experimentally measured the pure component spectra A_1 and A_2 are given by:

$$A_1 = \left(\frac{1}{1 - R_{\nu_2}} \right) A' - \left(\frac{R_{\nu_2}}{1 - R_{\nu_2}} \right) A \qquad (17)$$

$$A_2 = \left(\frac{1}{1 - R_{\nu_1}} \right) A' - \left(\frac{R_{\nu_2}}{1 - R_{\nu_1}} \right) A \qquad (18)$$

It is important to note that A_1 and A_2 are properly scaled to one another and as such can be used in the least squares fitting routines described later. This represents the internal calibration of the absorbance ratio method. The method can easily be extended to multicomponent systems and the case of overlapping absorption bands has also been considered.[16]

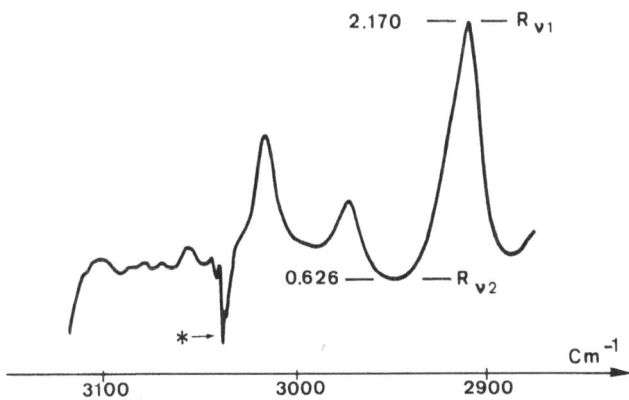

FIG. 8.   PET ratio spectrum for 3120–2850 cm^{-1} region; annealed at 160°C for two hours divided by unannealed. *Indicates artefact peak.[13]

As an example of this technique, consider the study of annealed PET. Factor analysis gave two components in the 2900–3100 cm^{-1} region. The ratio spectrum of PET annealed at 160°C for two hours divided by unannealed is shown in Fig. 8. The ratio coefficients R_{ν_1} and R_{ν_2} are measured to be 2·170 and 0·626 respectively. The concentration of the *trans* conformation for the unannealed sample was found to be 47·4%, the remainder being *gauche* ones. The film annealed for two hours at 160°C was found to contain 75·8% of *trans* and 24·2% of *gauche* conformations. Figure 9 illustrates the deconvoluted and internally scaled spectra of the *gauche* and *trans* conformations.

The same analysis has also been applied to PVC plasticised with dioctyl-phthalate (DOP) and quenched and annealed PVC.[16]

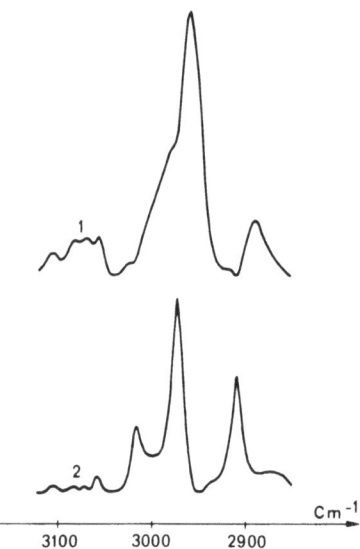

FIG. 9. PET C—H stretching region. (1) amorphous *gauche* conformation; (2) crystalline and amorphous *trans* conformation.[13]

3.3. Difference Spectroscopy

Besides the ratio method, digital subtraction of spectra appears to be an easy way to get an insight into polymer structure. This method was proposed some years ago[1] and up to now represents the most often used data processing system. The absorbance at any wavenumber ν for a spectrum can be written as:

$$A_1^\nu = A_{x_1}^\nu + A_{y_1}^\nu + A_{z_1}^\nu \qquad (19)$$

where A_1^ν is the total absorbance of all components at wavenumber ν; $A_{x_1}^\nu$ the absorbance of component x, $A_{y_1}^\nu$ the absorbance of component y and $A_{z_1}^\nu$ the absorbance of component z.

In the same way, for another spectrum in which the proportions of components x, y and z are different

$$A_2^\nu = A_{x_2}^\nu + A_{y_2}^\nu + A_{z_2}^\nu \qquad (20)$$

The subtraction is then obtained as:

$$A_S^\nu = A_1^\nu - kA_2^\nu = (A_{x_1}^\nu - kA_{x_2}^\nu) + (A_{y_1}^\nu - kA_{y_2}^\nu) + (A_{z_1}^\nu - kA_{z_2}^\nu) \qquad (21)$$

where k is an adjustable scaling parameter. Such an operation requires the Beer–Lambert law to be obeyed for the absorption bands that have to be subtracted ($A \leq 0.6$ absorbance unit). Besides this limitation, subjective evaluation of k is a potential source of error. In addition the previous treatment assumes that no interaction occurs between the different components of the mixture. Recent applications of this method will be reviewed in the next section.

3.4. Least-Squares Curve Fitting

Once the different component spectra of a mixture have been obtained, one may be interested in analysing other mixtures of the same components. This can be realised via the least-squares curve fitting method. In fact, curve resolving techniques are currently used to analyse infrared spectra. Usually these techniques involve fitting Gaussian or Lorentzian curves by a least-square criterion to spectra consisting of many overlapping absorption bands[18] in order to determine the number and frequency of each absorption band contributing to the spectrum. From a practical point of view, absorption bands of polymers are usually neither Gaussian nor Lorentzian and Koenig et al.[19] proposed a least-square computing algorithm for curve fitting of infrared absorbance data. The fitting equation employed is as follows:

$$\sum_{i=1}^{n} A_{i,k} = \sum_{j=1}^{m} \left(\sum_{i=1}^{n} W_i A_{i,j} A_{i,k} \right) X_j \tag{22}$$

and
$$S_i = \sum_{j=1}^{m} X_j A_{i,j}$$

where n is the number of spectral elements (data points) in each spectrum and m the number of basis spectra used in the fitting procedure. $A_{i,k}$ represents the absorbance data for the ith spectral element of the kth basis spectrum; S_i is the data value in ith channel of the composite spectrum; $W_i = 1/S_i$ is a weighting factor and X_i is the number which, when multiplied by the appropriate basis spectrum and summed for all components best fits the experimental spectrum. The feasibility of the method was demonstrated by an analysis of o-, m- and p-xylene mixtures.[19]

The analysis of compatible blends of isotactic polystyrene (iPS) and poly(2,6-dimethyl 1,4-phenylene oxide) (PDMPO) is a good example of the application of the method to polymer systems.[19] As shown in Fig. 10 a good least-square fit of iPS and PDMPO blend in a 50/50 weight ratio is obtained in the 3200–2800 cm^{-1} region. After the spectrum of the blend was stripped of its components a residual indicative of interactions between the components remains as illustrated in Fig. 11. Similar inter-

106 B. JASSE

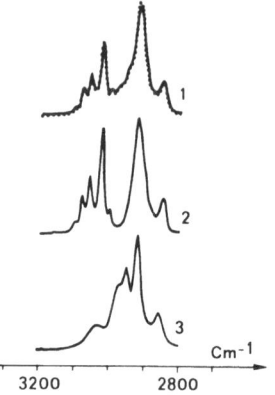

FIG. 10. Least-square curve fitting of iPS–PDMPO polyblend for 3200–2800 cm^{-1} region: (1) 50:50 weight ratio iPS/PDMPO polyblend (solid line); best fit of pure iPS plus pure PDMPO (●); (2) iPS; (3) PDMPO.[19]

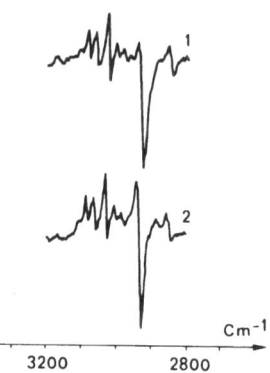

FIG. 11. Interaction spectra from iPS/PDMPO polyblends: (1) 50:50 iPS/PDMPO polyblend spectrum stripped of both pure components; (2) 60:40 iPS/PDMPO polyblend spectrum stripped of both pure components.[19]

action spectra are obtained for a wide range of composition of the blends suggesting that this kind of interaction is characteristic of iPS and PDMPO. Such an analysis is however based on the assumption that the least-squares coefficients are not significantly affected by the variation of a single absorption band. Although such effects could lead to errors in determining the composition of a polymeric system, the use of iPS and PDMPO reference spectra scaled such that the curve-fit program calculates a 50/50 percentage composition for a standard 50/50 blend gives a quite good agreement with the actual composition of different iPS/PDMPO blends as shown in Table 1.

TABLE 1
LEAST-SQUARES CURVE FITTING OF iPS–PDMPO BLENDS[19]

Known wt% composition		Predicted by curve-fitting	
% iPS	% PDMPO	%iPS	%PDMPO
50·0 ± 1	50·0 ± 1	50·00 ± 0.61	50·00 ± 0.97[a]
40·0 ± 1	60·0 ± 1	36·07 ± 1·02	63·93 ± 1·62
60·0 ± 1	40·0 ± 1	60·41 ± 0·42	39·59 ± 0·66
90·0 ± 1	10·0 ± 1	90·37 ± 0·30	9·63 ± 0·47

[a] Calibration.

4. RECENT APPLICATIONS OF FTIR SEPARATIONS THROUGH DATA PROCESSING

Besides the previous examples illustrating the different techniques of data processing numerous papers have recently been published pointing out the tremendous development of FTIR spectroscopy in the study of polymers.

4.1. Structural and Conformational Changes in Polymers

Polymer properties strongly depend on the structure and conformations of the chains, and the absorbance subtraction technique allows, for the first time, the determination of the spectra of each of the pure species. A recent application involves the study of conformational change of the methylene sequence induced in poly(tetramethylene terephthalate) (PBT) by thermal or mechanical treatment.[20, 21] This polymer undergoes a reversible phase transition dependent on the sample treatment. The α form obtained by annealing an amorphous sample is attributed to an approximately *gauche–trans–gauche* conformation of the aliphatic chain segments. Under stress the α form transforms to an extended *trans–trans–trans* conformation (β form). A schematic description of this phase transition induced by stretching, relaxation and annealing is given in Fig. 12. Characteristic

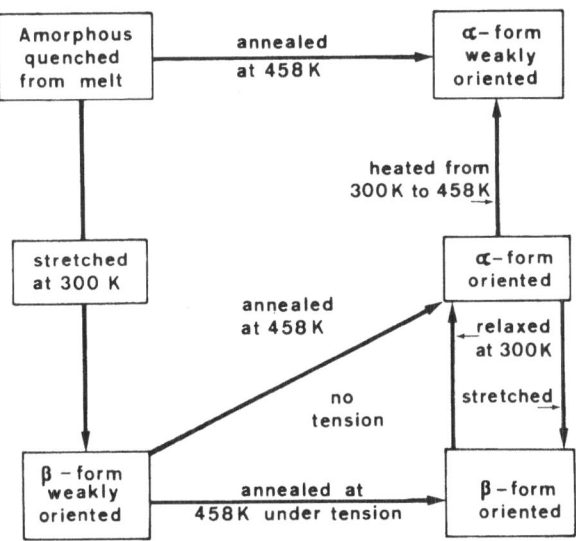

FIG. 12. Sample pretreatment of poly(tetramethylene terephthalate) (PBT).[21]

spectral changes have been observed during these different treatments and typical absorption bands have been assigned to the α and β forms on the basis of spectral changes and a normal coordinate analysis of the β form[21] (see Table 2). Appropriate subtractions gave the spectra of α and β forms shown in Fig. 13.

TABLE 2

ASSIGNMENT OF THE CONFORMATIONALLY SENSITIVE ABSORPTION BANDS OF PBT[21]

α form	β form	Assignment
1 460	1 485	CH$_2$ bending
1 386	1 393	CH$_2$ wagging
918	960	Coupled C—O stretch, C—O—C—, O—C—C bend

Conformational changes in amorphous poly(ethylene terephthalate) films during volume recovery at 50°C have been investigated by FTIR.[22] Difference spectra between quenched and annealed films indicate that the fraction of the *gauche* conformation of the methylene group increases in an orderly manner and that the out-of-plane motion bands of the benzene ring decrease in intensity at or beyond annealing times corresponding to

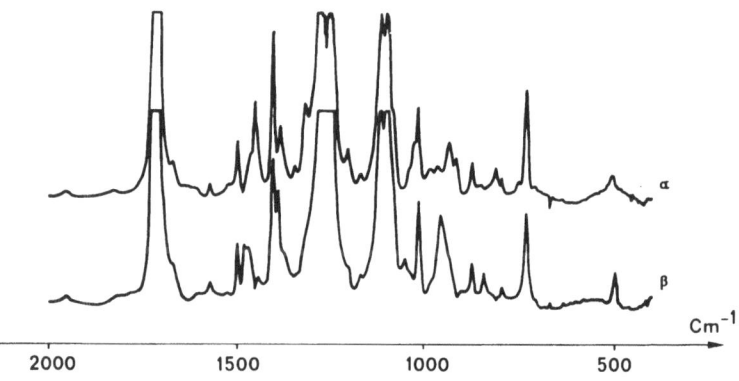

FIG. 13.   Spectra of α and β forms of poly(tetramethylene terephthalate) (PBT).[21]

embrittlement. The authors suggest that such measurements could be able to provide understanding of embrittlement on a molecular basis.

Another polymer, poly(vinylidene fluoride) (PVF$_2$) has received considerable attention because it exhibits piezoelectric and pyroelectric properties. Two types of chain conformations are found to occur: phase I, having an all *trans* structure and phase II having a *TGTG'* chain structure. A third form of PVF$_2$ is observed when the polymer is crystallised under specific conditions. A FTIR study[23] was performed in order to obtain information on the structure of form III. Cast films from a 9% (by volume) solution in *N-N'*-dimethylacetamide (DMA) were treated at 200°C and then air quenched to room temperature. This treatment resulted in films of unoriented phase II which could be transformed to phase I by biaxially drawing at room temperature. Samples of phase II were obtained by casting films from a 6·3% (by volume) solution in DMA at 60°C, subsequent heating at 200°C and then air quenching. Samples of phase III were prepared by casting films from a 6·3% solution in DMA at 60°C. Crystalline spectra were obtained from two samples with different degrees of crystallinity. The criteria for subtraction was to remove the contribution from the amorphous regions of the films. The crystalline spectrum of phase III is not consistent with current models for the crystal structure. Several chain conformations are acceptable from the spectral point of view but the *TTTGTTTG'* conformation is consistent with the X-ray results and theoretical calculation.

As far as *trans*-1,4 polyisoprene (tPI) is concerned, two polymorphic forms exist.[24] In an as cast film from a chloroform solution the crystalline component is predominently in the α form. The β form is obtained by quenching at −25°C a film in the amorphous state at 80°C. The spectra of α and β forms are obtained by subtracting the amorphous spectrum of tPI and are shown in Fig. 14. Band assignments were made using normal coordinate calculations on molecular models of the two polymorphic forms of tPI and a good agreement was obtained with experimental values.

The structure of the gel form of isotactic polystyrene (iPS) obtained by high supercooling conditions remains controversial.[25] The gel forming micelles should have a different chain conformation from the *trans–gauche* 3$_1$ helical structure usually observed in this polymer. Different methods of measurement lead to contradictory results and interpretations. The vibrational spectrum of iPS is known to be very sensitive to conformation[26] and it was expected that the data handling capabilities of FTIR would allow information on the conformation or configurations involved in this structure to be obtained. Painter *et al.*[25] studied the 800–1400 cm^{-1} region of the iPS spectrum which is conformationally sensitive. Decalin was removed for iPS gel by digital subtraction. The 'gel crystallites' spectrum is compared to crystalline iPS and atactic PS in Fig. 15. The out-of-plane bending modes of the ring at 922 and 842 cm^{-1} in the spectrum of the crystal form are shifted to 916 and 849 cm^{-1} in the spectrum of the gel. Important spectral changes are also observed in the infrared bands grouped near 1300, 1200, and between 1050 and 1085 cm^{-1} respectively. These results indicate the presence of a conformation which is different from the 3_1 helix and a nearly extended structure is strongly indicated.

A decrease in the amount of *trans* conformation was observed with increase in temperature and pressure of vitrification in the conformational recovery of poly(vinylchloride) (PVC) glasses prepared under pressure.[27] Changes in absorbance of conformationally sensitive absorption bands were associated with conformational re-equilibration occurring above the glass temperature of PVC (80°C).

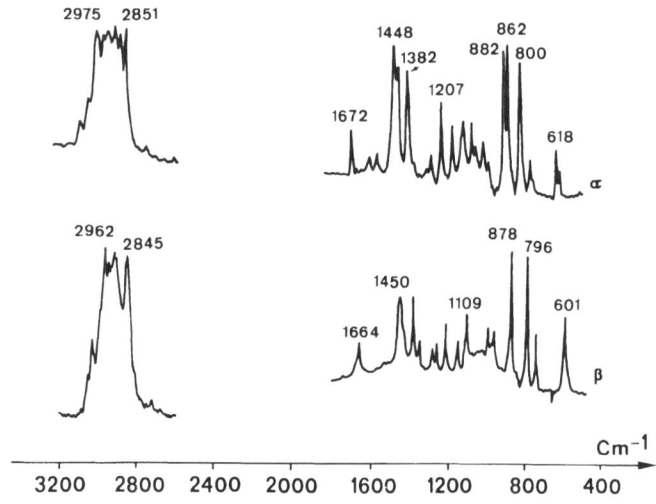

FIG. 14.   Spectra of α and β forms of *trans*-1,4 polyisoprene.[24]

A recent FTIR study of conformational defects in crystalline poly-ethylene (PE)[28] gives an interesting example of the use of a small amount of deuterated material dispersed in a similar hydrogenated polymer matrix as a tool to getting an insight into the molecular behaviour of polymers. The CD_2 group in low concentration (*ca* 5%) displays absorption bands which are directly associated with the conformational structure of adjacent pairs of C—C bonds. More precisely, the rocking mode of the CD_2 group in *trans* sequences is located around 620 cm^{-1} while the same vibration is observed between 646–651 cm^{-1} in *trans–gauche* structures. Both these bands slightly overlap the shoulder of the 720 cm^{-1} rocking mode of the CH_2

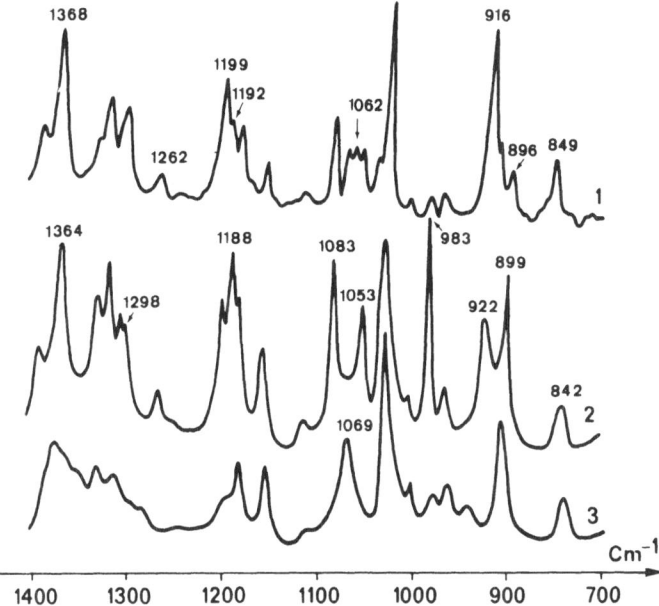

FIG. 15. Comparison of the infrared spectrum of the ordered gel component of iPS, the 3_1 helical crystal and atactic polystyrene: (1) gel; (2) crystal form; (3) atactic polystyrene.[25]

group. Spectra were measured between 25°C and 120°C and difference spectra indicate an increase of the *trans–gauche* conformation in crystalline PE with an increase of temperature. The increase begins at about 80°C and accelerates above 100°C. Quantitative measurements are widely discussed on a baseline choice basis.

FTIR spectroscopy has also been used to study the extent of interchain hydrogen bonding and its temperature dependence in a segmented polyurethane.[29]

4.2. Polymer Blends

The interest in polymer blends as a way to meet new market applications with minimum development cost has rapidly increased over the last decade. Usually two different polymers will not mix on the segmental level but an especially favourable interaction between two polymers can allow one to obtain homogeneous blends. FTIR is a potential tool for the investigation of the mutual compatibility of various polymers: the very small spectral changes which are introduced as a result of the interactions can be detected by this method. If two polymers are incompatible, i.e. phase separation occurs, one can synthesise a spectrum of the blend by coadding in the appropriate proportions the spectra of the two pure components which will compare with the observed spectrum of the blend. On the other hand, when polymers are compatible, only one phase exists and there should be observable spectral differences between the coadded spectra of the pure components and the spectra of the mixture. Figure 16 shows the spectra of pure poly(methyl methacrylate) (PMMA), pure

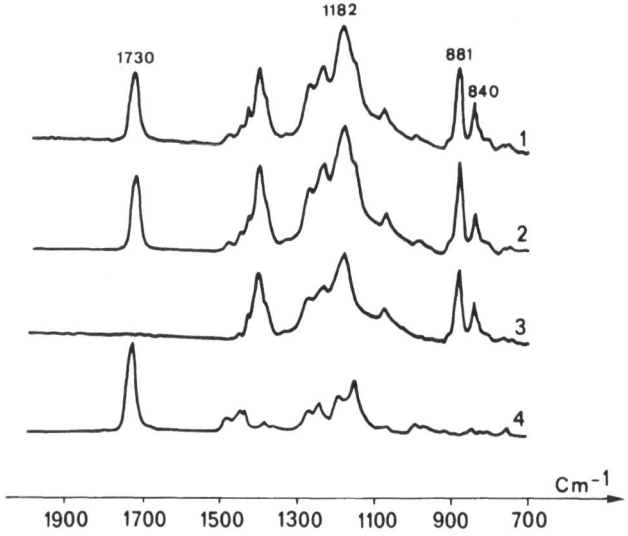

FIG. 16. PVF$_2$/PMMA blend of 75:25 parts by weight: (1) Coadded spectrum 3 + 4; (2) PVF$_2$/PMMA blend spectrum; (3) PVF$_2$ spectrum; (4) PMMA spectrum.

PVF$_2$, and a blend of PVF$_2$/PMMA containing 75:25 parts by weight, in the 700–2000 cm^{-1} region. The synthesised spectrum is almost identical to the experimental spectrum of the blend. Hence this blend is incompatible. Figure 17 shows the results of an identical experiment using a PVF$_2$/PMMA blend composition of 39:61 parts by weight. This blend is considered compatible and it is impossible to approximate the experimental spectrum by absorbance addition of the two polymer spectra.[30] In particular, absorption bands at 1180, 1072, and 840 cm^{-1} are not matched.

Compatible and incompatible polyester–PVC blends have also been considered. Two systems have been studied: poly(ϵ-caprolactone) (PCL)/

PVC blends[31] which are compatible in the melt and exhibit partial compatibility in the solid state (above 60% of PVC) and poly(β-propiolactone) (PPL)/PVC blends[32] which are known to be incompatible.

Figure 18 shows the infrared spectra of the carbonyl stretching vibration (in the range 1600–1800 cm^{-1}) for the different blends. Specific interactions between PCL and PVC are clearly indicated. In the molten state, the 1737 cm^{-1} band in PCL shifts to 1731 cm^{-1} in the 5:1 molar blend (Fig. 18a). In the solid state (Fig. 18b) the spectrum of pure PCL indicates the presence of crystalline (1724 cm^{-1}) and amorphous (1737 cm^{-1}) bands. At mole

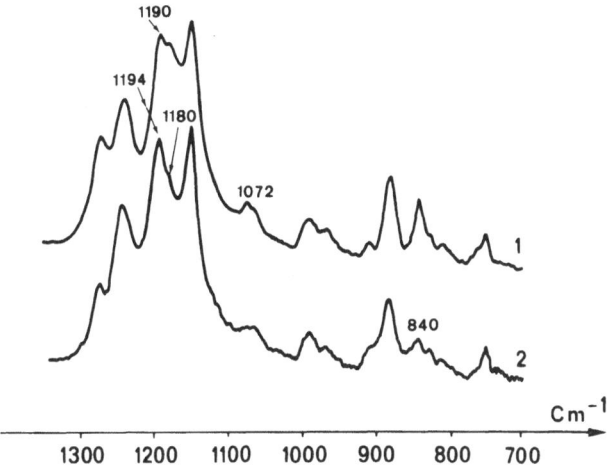

FIG. 17. PVF$_2$/PMMA blend of 39:61 parts by weight: (1) Coadded spectrum; (2) PVF$_2$/PMMA blend spectrum.

ratios up to 2:1 of PVC to PCL, the spectra indicate that the blends consist, in the solid state, of a crystalline and amorphous phase. As the PVC concentration increases, a parallel increase of the intensity of the amorphous band is observed. Moreover, the frequency shifts observed for both the crystalline and amorphous bands as a function of the composition of the blend suggest specific interactions between the two polymers occur.

No shift is observed in the carbonyl stretching vibration of PPL/PVC blends, in the molten state or in the solid state, over the entire range of compositions (Fig. 18c and d) and the two polymers are incompatible.

Similar shifts of the asymmetric ether stretching vibration have been observed in compatible PDMPO/PS blends.[33]

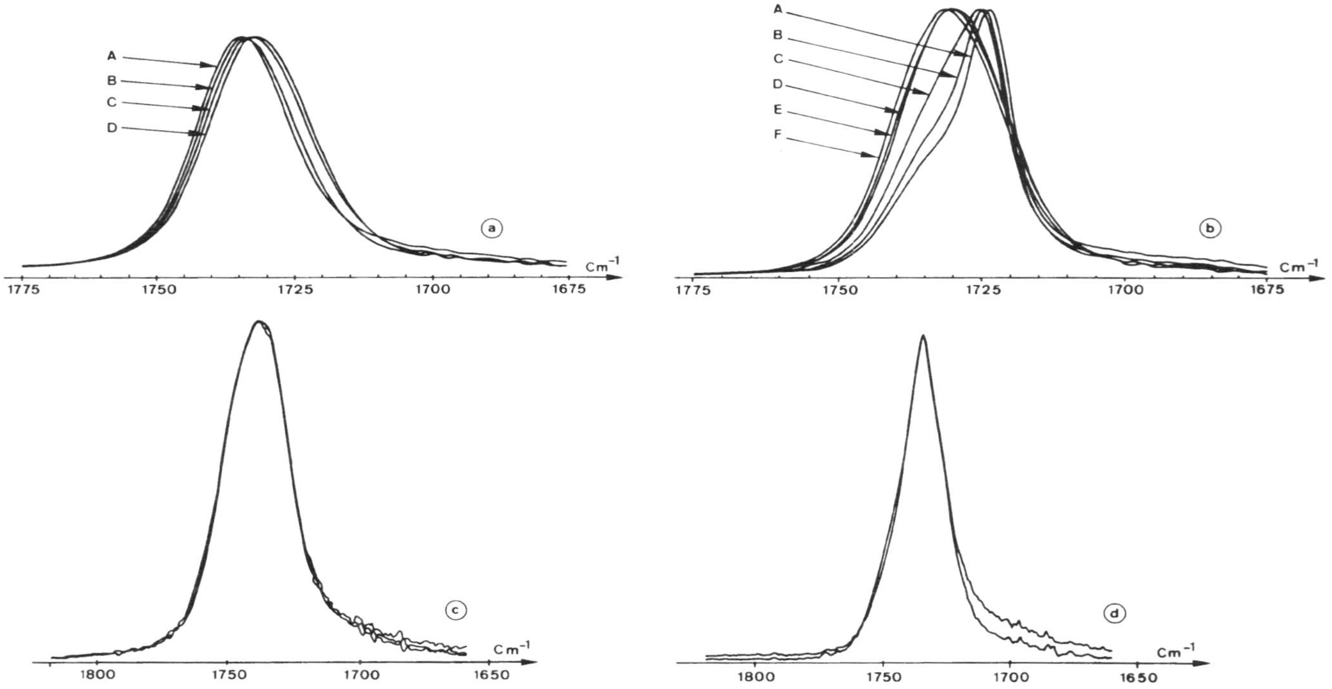

FIG. 18. Polyesters/PVC blends: (a) spectra of PVC/PCL blends recorded at 75° for (A) pure PCL, (B) 1:1, (C) 3:1 and (D) 5:1 molar PVC:PCL, respectively; (b) spectra of PVC/PCL blends recorded at room temperature for (A) pure PCL, (B) 1:1, (C) 2:1, (D) 3:1, (E) 5:1, and (F) 10:1 molar PVC:PCL, respectively; (c) spectra of PPL, PPL/PVC (55:45 molar) and PPL/PVC (20:80 molar) recorded at 80°C; (d) spectra of PPL, and PPL/PVC (30:70 molar) recorded at room temperature.[31, 32]

4.3. Chemical Transformation and Degradation in Polymers

Degradation in polymers is an irreversible process in which physical properties undergo injurious changes. An understanding of the mechanism of degradation may lead to a new approach to stabilisation. Usually degradation induces small changes in polymer spectra and absorbance subtraction appears to be an interesting way of examining the reacting portions of chains and the reaction products.

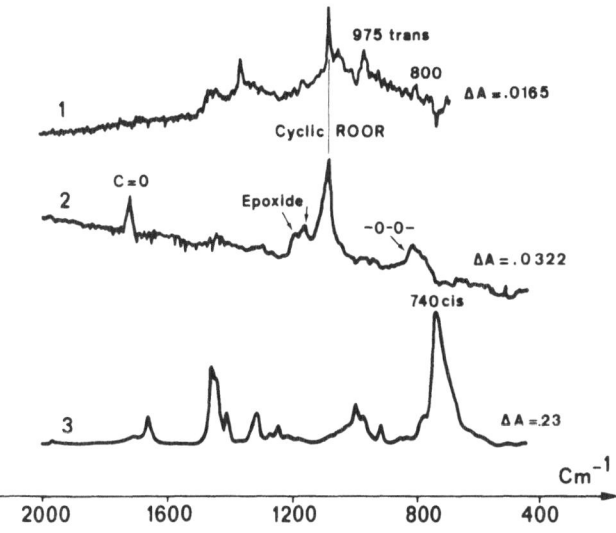

FIG. 19. Oxidative degradation of *cis*-1,4 polybutadiene (BR). Difference spectra obtained under different conditions of oxidation: (1) BR + di-tert-butylsulphoxide (5 wt%) oxidised 3 days at 25°C and (2) BR oxidised one month at 0°C, minus unoxidised BR; (3) unoxidised.

A first example of such an approach is given by the oxidative degradation of *cis*-1,4 polybutadiene (BR) under different conditions.[34] In Fig. 19, the spectrum of unoxidised BR is compared with two difference spectra of oxidised BR minus unoxidised in the range 2000–450 cm^{-1}. Oxidation at 0°C for one month in the absence of light and moisture results in the appearance of different absorption bands which have been identified as peroxides and epoxides. Oxidation at higher temperature induces the formation of *trans* methine. *Cis*-1,4-polyisoprene and *trans*-1,4-polychloroprene degradation mechanisms are also described in the same paper.

Cured epoxy resins degradation has been investigated under different conditions.[35] An order of stability of the functional groups was obtained and the degradation processes related to the classical autocatalytical oxidation of aliphatic hydrocarbon segments. FTIR absorbance subtraction of virgin poly(tetrafluoroethylene) (PTFE) and γ-ray irradiated samples indicates a decrease in crystallinity and an increasing concentration of free and bonded —COOH end groups consistent with chain scission.[36]

Polysiloxanes are often used at surfaces to enhance adhesion in glass fibre reinforced composites. FTIR spectroscopy brought experimental evidence of chemical bonding of the coupling agent to the glass surface.[37] A recent study is devoted to the hydrothermal degradation of various coupling agents on E-glass fibres.[38]

4.4. Stress Induced Changes in Polymers

The infrared vibrational frequencies of bonds in a molecule depend on the force constants (determined by the strength of atomic bonds) and bond angles. Subtle but definite changes of infrared absorption frequencies and intensities appear in polymers in the stressed state (Fig. 20). Such modifications give information about molecular deformation processes and load bearing abilities of single chains. A review was recently published on this subject.[39] Pioneer works were performed on dispersive instruments but the minute character of these spectral changes suggests FTIR as a tool well suited to such studies.

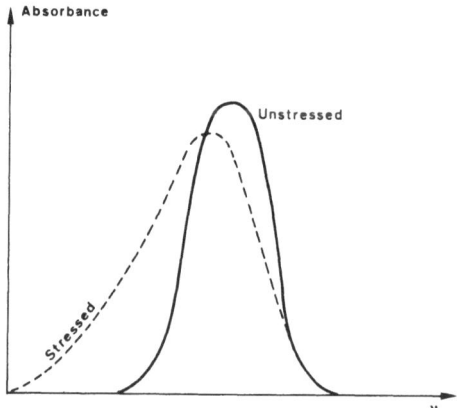

FIG. 20. Schematic of typically observed infrared band deformation and frequency shifts of chain vibrations in stressed polymers.[39]

Studies of stress-induced changes in poly(ethylene terephthalate) (PET)[40] pointed out that the inter-microfibrillar chains are the first to be stressed followed by stress transmission to the crystalline regions. These stresses induce stretching of the chain bonds, conformational changes from *gauche* to *trans* of ethyleneglycol linkage and modification of interchain interactions. Changes occurring at the molecular level in stressed polystyrene have been studied by subtracting a stressed from an unstressed spectrum of this polymer.[41] The result is illustrated in Fig. 21.

One can observe additional absorption bands at 889, 1141, 1400 and 1506 cm^{-1}; bands at 540, 1448, 1492 and 1602 cm^{-1} exhibit splitting. All these changes of absorption bands are related to various vibrational modes

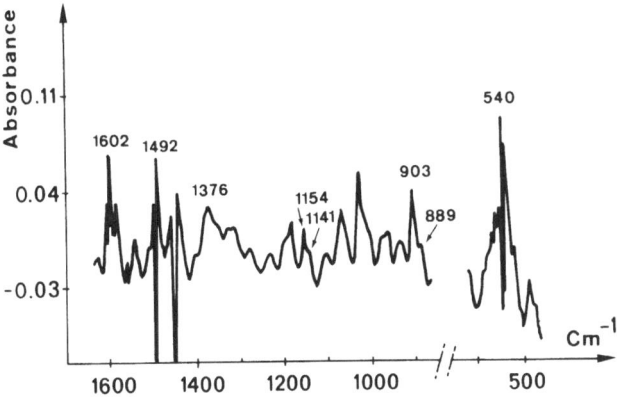

FIG. 21.   Infrared spectrum of (stressed minus unstressed) atactic polystyrene film.[41]

of the benzene ring and it is suggested that phenyl groups on adjacent chains interact extensively under stress, changing the electron distribution on the benzene ring. A similar work has been published relative to polycarbonate.[42] A small frequency shift was observed for the 556, 1015 and 1080 cm^{-1} absorption bands.

4.5. Surface Studies by ATR

Multiple internal reflection spectroscopy is a very interesting tool for analysing surface properties. FTIR provides advantages in this field on account of the improved signal-to-noise ratio and higher energy throughput. This method is particularly attractive for comparing crystallinity and molecular orientation between surface and bulk of oriented polymers. Tse and Paik Sung[43] proposed a rotatable sample holder allowing the polymer film to be mounted and rotated orthogonally while maintaining identical contact necessary to obtain reliable data. The results obtained by these authors indicate a lower orientation of the surface of cold drawn polyethylene as compared with the bulk while similar orientations were observed when comparing the surface and the bulk of the cold drawn polypropylene.

Surface and second layer spectra of surface coated poly(vinyl alcohol) and polypropylene films have been obtained using an ATR device at two incident angles and digital subtraction.[44]

5. FAST-SCANNING FTIR SPECTROSCOPY

The availability of fast-scanning FTIR spectrometers allows the study of short time phenomena which cannot be studied by dispersive infrared spectrometers. Contrary to conventional instruments where only registration of small frequency regions or with constant wavenumber can be obtained, FTIR spectroscopy gives a series of complete spectra representative of the evolution of the system under study. The observed differences in the spectra as a function of time can be related to changes in the structure or composition of the material (e.g. orientation or relaxation of polymer chains under stress; degree of crystallinity; changes of conformational structure; progress in polymerisation or degradation reactions, etc.).

5.1. Studies of Polymers during Elongation

The advantages of a FTIR-system for the study of orientation behaviour during the application of an external stress have been extensively exploited for two years. For this purpose, special stretching machines have been built[7, 45, 46] which allow a polymeric film to be uniaxially stretched in the infrared beam of the spectrometer. It is possible to obtain simultaneously the stress–strain diagram and the infrared spectra of the sample. For orientation measurements, a rotatable polariser allows one to fix the polarisation direction alternatively parallel or perpendicular to the stretching direction. The dichroic ratio $R = A_{\parallel}/A_{\perp}$ of any absorption band at small strain intervals may then be obtained.

Different studies on structural changes and orientation processes occurring in polyethylene (PE) during a stress–strain experiment have been recently published.[5-7, 45-49] The changes in PE films during elongation are mainly observed on the $730/720 \, \text{cm}^{-1}$ (CH_2 rocking) absorption bands.

The transition moment vectors of the corresponding vibrations are parallel to the *a* and *b* axes of the crystalline cell, respectively. The samples used were blown PE films having an initial orientation: unit cell *c* and *a* axes possess a preferred orientation perpendicular to the film plane and parallel to the machine direction (which is identical to the later stretching direction) respectively. As far as structural changes during the stretching process are concerned, the orthorhombic structure of crystalline PE is

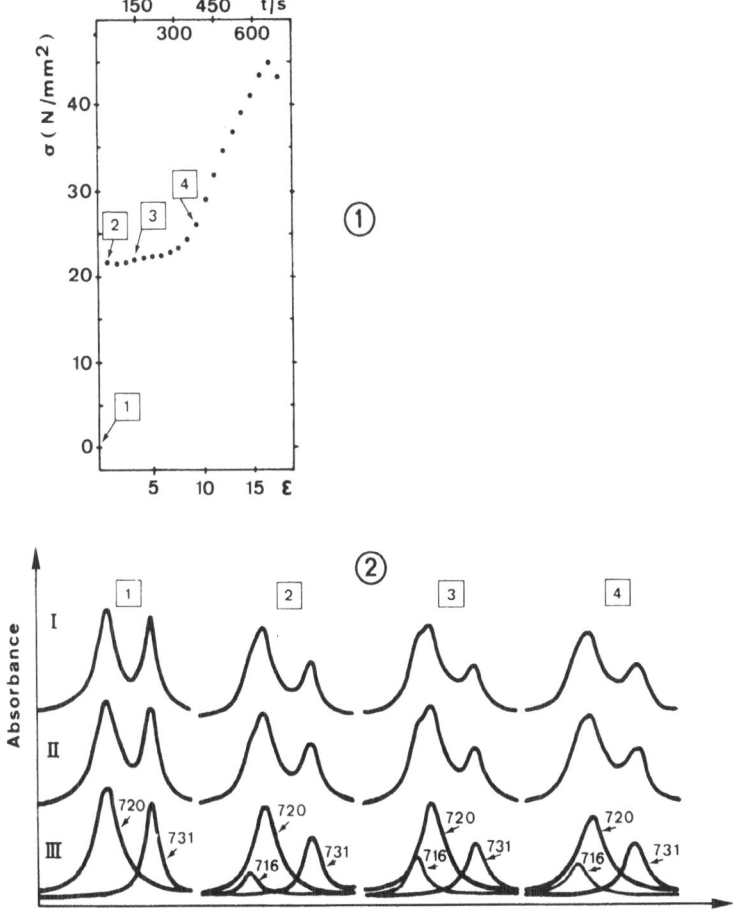

FIG. 22.   (1) Strain–stress diagram of a polyethylene film. (2) Experimental (row I), synthesized (row II) and resolved (row III) bands of the CH_2-rocking mode. The experimental spectra were scanned at the indicated positions (squared numbers) of the strain–stress diagram.[48]

partially transformed into a monoclinic structure as demonstrated by the appearance of a characteristic absorption band at 716 cm^{-1}.[45, 46, 48] Figure 22 shows the changes observed in the 700–800 cm^{-1} region of the infrared spectra as the stretching process develops. Faster stretching velocities induce an increase in the amount of the monoclinic structure.[49]

Besides structural changes, it is interesting to follow the orientation process during stretching. This was achieved by the study of dichroic ratio

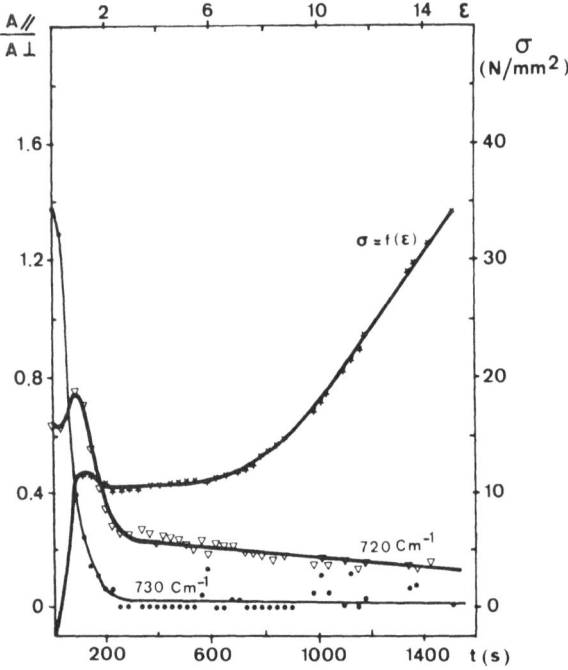

FIG. 23.   Change of dichroic ratio as a function of strain for the CH$_2$-rocking vibrations of a polyethylene film.[49]

($R = A_{\parallel}/A$) change of the 720 cm^{-1} and 730 cm^{-1} absorption bands as a function of strain. Figure 23 illustrates the change of dichroic ratio as a function of strain or time. For small strains ($\epsilon < 0.5$) original orientation increases slightly. Characteristic changes in dichroic ratios then occur with the beginning of necking (around $\epsilon \simeq 1$). In the interval $1 < \epsilon < 3$, a rotation of a and b axes in directions perpendicular to the stretching direction occurs (c axis rotates into the stretching direction). Simultaneously, the intensity of the 730 cm^{-1} absorption band decreases as compared to the 720 cm^{-1} absorption band. This behaviour is related to the beginning of the disappearance of the orthorhombic structure of the unit cell by unfolding of the lamellae.[49] Similar results were obtained when scanning spectra through the necking zone.[49]

As previously discussed, poly(tetramethylene terephthalate) (PBT) undergoes conformational changes when subjected to deformation or annealing. Fast-scanning FTIR spectroscopy allows the study of the time

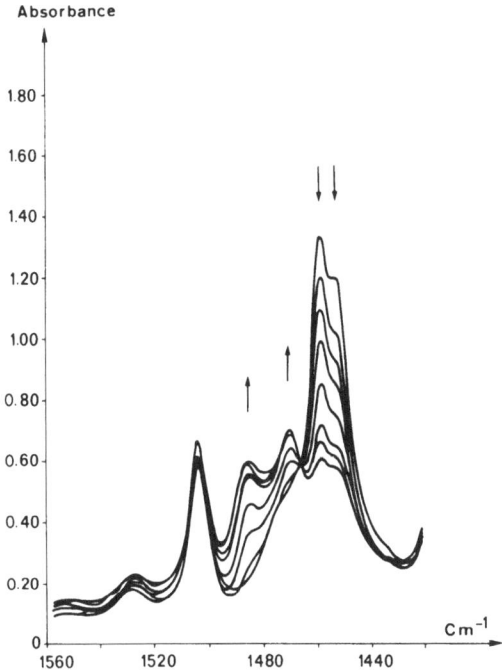

FIG. 24.   Isosbestic point ($1466\ cm^{-1}$) observed in the $\delta(CH_2)$ absorption region of infrared spectra recorded during loading of a PBT film.[50]

dependence of these phenomena. Siesler[45, 50, 51] submitted PBT to loading–unloading cycles and obtained dynamic spectra at small strain intervals. An isosbestic point is observed at $1466\ cm^{-1}$ (see Fig. 24) as a consequence of the decrease and increase of the $1460/1450\ cm^{-1}$ and $1485/1470\ cm^{-1}$ [$\delta(CH_2)$] absorption bands of the unstressed and stressed structural units respectively. As shown in Fig. 25 an important change of absorption intensity of the $\delta(CH_2)$ absorption bands located at 1460 and $1485\ cm^{-1}$ is observed in the 5–15% strain region but no significant change occurs during stress relaxation, the sample being held at a 12·5% constant strain.

From these results, slippage of intercrystalline segments is suggested as the most probable mechanism of stress relaxation.

The temperature effect on conformational changes in PBT has been studied by Holland-Moritz and co-workers.[47] Spectra were recorded in the 700–1050 cm^{-1} region at 20 s time intervals during rapid heating from 338 K to 468 K. A similar analysis of polypropylene orientation behaviour[52] has shown a drastic change of the dichroic ratio of 1378, 999 and 975 cm^{-1} absorption bands during necking, indicative of a preferential perpendicular alignment of the CH$_3$ side groups and of a parallel alignment of the helix axes of the polymer with respect to the stretching direction. In this

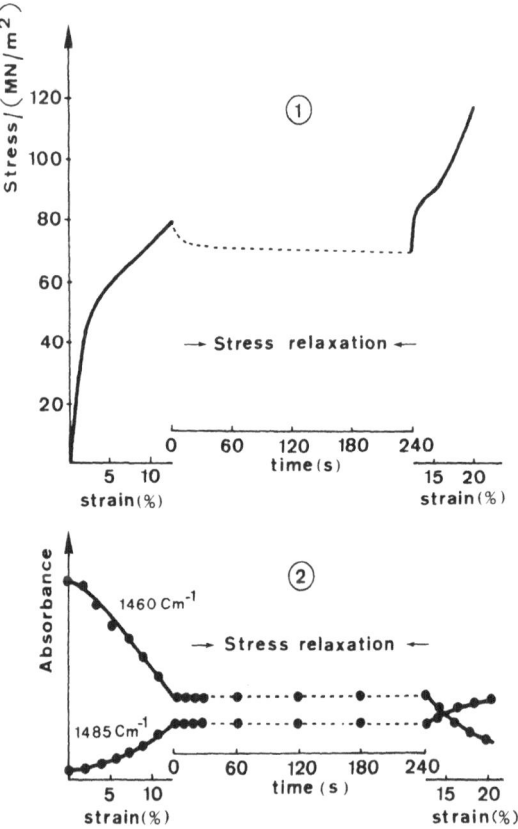

FIG. 25. Elongation and stress relaxation of a PBT film: (1) stress–strain/time diagram; (2) absorbance variations of the 1460 and 1485 cm^{-1} δ(CH$_2$) absorption bands during elongation and stress relaxation.[51]

stage of deformation, orientation processes involve translation and rotation of the entire polymer chains. In the strain hardening region, further orientation is preferentially observed for conformationally regular segments.

Segmental orientation of polyesterurethane films during uniaxial deformation has also been investigated in the same way.[6, 7]

5.2. Chemical Reactions Studies

Infrared spectroscopy is widely used in order to obtain information on polymer reactions and kinetics of polymerisation.[7] Spectral changes provide a good method of following the polymerisation processes or identifying new functional groups created during any chemical reaction on a polymer.

As an example we will consider the formation of polyurethane foams.[53] This study was performed using a special ATR cell allowing characterisation of changes of functional groups in short time intervals. The polyurethane foam formation is illustrated in Fig. 26. The three-dimensional representation clearly indicates the disappearance of isocyanate group (decrease of the ν—N=C=O absorption band at 2275 cm^{-1}) and the competitive formation of urethane and urea linkages (appearance of the ν C=O absorption band at 1725 cm^{-1} and of the amide I and amide II bands between 1500 and 1600 cm^{-1}).

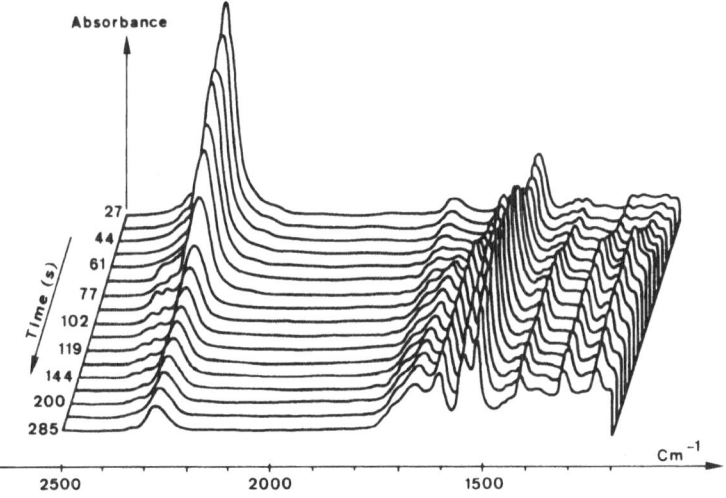

FIG. 26.   FTIR spectroscopic study of polyurethane foam formation.[53]

Fast-scanning FTIR spectroscopy has also proved to be very useful in realising real-time on-line analysis of gaseous or liquid eluates from a gas chromatograph or a liquid chromatograph respectively. Problems such as combustion or degradation mechanism of polymers or identification of polymer mixtures by gel permeation chromatography can be solved in this way. This field is just beginning to be explored and is discussed in detail in references 6, 7 and 47.

6. FTIR–PHOTOACOUSTIC SPECTROSCOPY

Recently developed FTIR–photoacoustic spectroscopy appears to be an attractive method of obtaining spectra of optically dense samples without sample preparation. Such samples as pellets, chips, rough surfaces, coatings, catalysts, etc., can be examined in the infrared range when

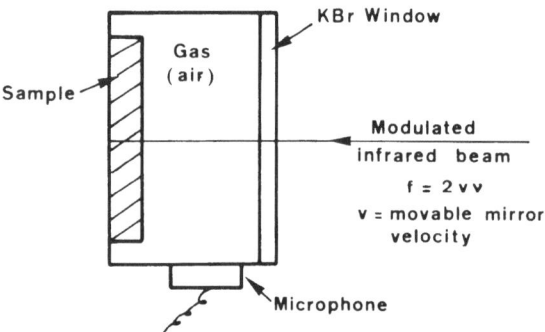

FIG. 27.   Scheme of a photoacoustic cell.

conventional techniques are inconvenient. For this purpose, the conventional detector of the FTIR spectrometer is replaced by a photoacoustic (PA) cell, the scheme of which is given in Fig. 27. The basic principle of the PA cell relies on the fact that in an absorption band the incoming energy is essentially converted into heat. The periodic change of the temperature of the sample induces a periodic change of gas pressure in the cell. These pressure changes, measured by the microphone, form the PA signal. A theoretical treatment of PA effect may be found in references 54–56. An inherent problem with PA spectroscopy is that it is a detector-noise limited technique. For such experimental conditions, it is well known that a FTIR spectrometer should be very efficient in improving the signal-to-noise ratio, according to the multiplexing advantage (Fellgett's advantage) and an increased light throughput. Rockley,[57, 58] Vidrine[59] and Royce et al.[60-62] proved the feasibility of the method in the 4000–400 cm^{-1} infrared range using commercially available FTIR spectrometers. The effective modulation frequency f of the infrared beam at a given wavenumber ν is $f = 2v\nu$ where v is the linear velocity of the moving mirror of the interferometer. The depth of the sample studied depends on light modulation frequency which can be varied by changing the interferometer mirror velocity. Saturation of the PA signal can thus be suppressed. It is necessary to normalise the PA spectrum obtained and to take into account both the

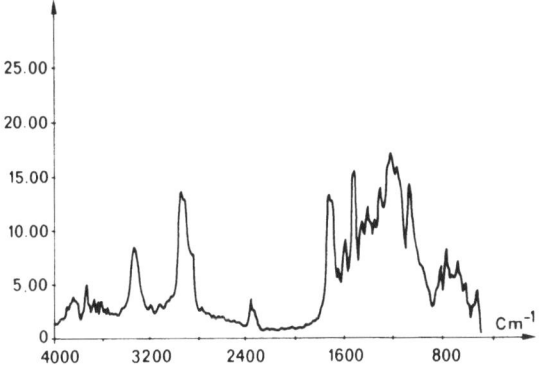

FIG. 28.   Photoacoustic spectrum of polyurethane feedstock. Mirror velocity is 0·74 cm s^{-1}; y axis is percent signal relative to carbon black.[59]

characteristics of the PA cell and instrument function. This is usually realised by taking a carbon black reference spectrum.

Compared to transmission spectroscopy the PA technique requires much greater measurement times which are expressed in minutes, not seconds. For a material with strong absorption bands, as few as 32 scans (≈ 2 min) will give a spectrum good enough for a rough identification of the material. In order to improve the signal-to-noise ratio, about 1000 to 2000 scans are usually made with a resolution of 4 to 8 cm^{-1}. Figure 28 shows the PA spectrum of the polyurethane chip recorded with an 8 cm^{-1} resolution, 1024 scans and a mirror velocity of 0·74 cm s^{-1} (measuring time $\approx$ 30 min). As shown in Fig. 29 similar spectra of polystyrene were obtained using FTIR and FTIR–PA spectroscopy. One can notice the absence of interference fringes in the PA spectrum.

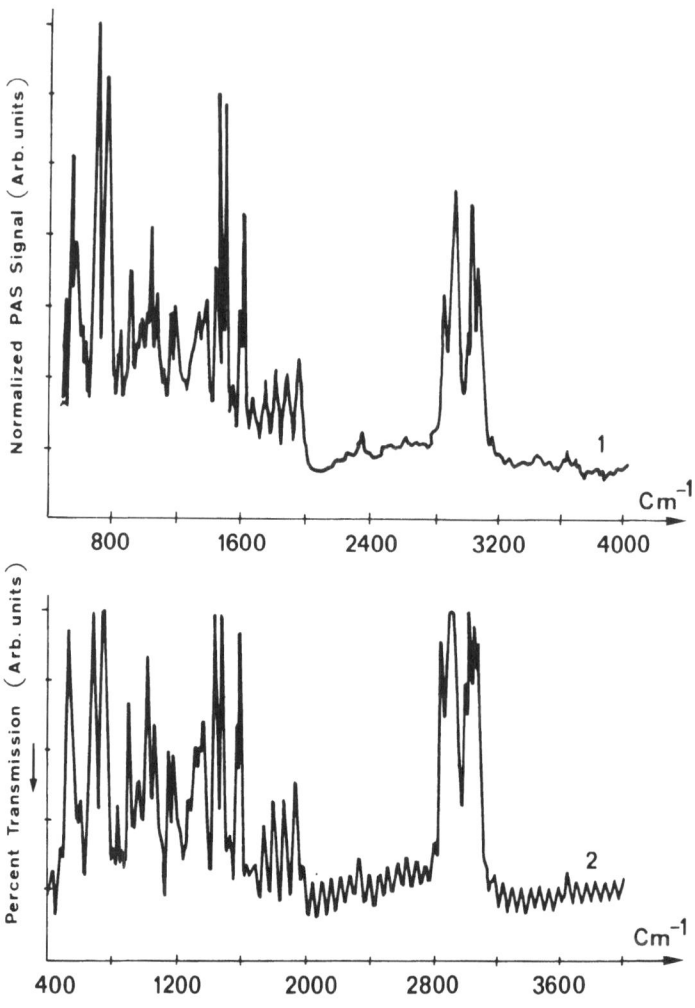

FIG. 29.   Comparison of photoacoustic and transmission spectra of polystyrene:
(1) PA spectrum; (2) transmission spectrum.[62]

The FTIR–PA technique is still in the early stage of development and requires further research to reach the quantitative level. Measurement times are much longer than those required for usual FTIR spectroscopy and this technique should find a major development in the study of samples which cannot be studied easily by other infrared techniques.

ACKNOWLEDGEMENT

The author wishes to thank Prof. J. L. Koenig for his advice and assistance.

REFERENCES

1. KOENIG, J. L., *Appl. Spectrosc.*, 1975, **29**, 293.
2. COLEMAN, M. M. and PAINTER, P. C., in: *Applications of Polymer Spectroscopy*, E. G. Brame, Ed., 1978, Academic Press, New York, p. 135.
3. D'ESPOSITO, L. and KOENIG, J. L., in: *Fourier Transform Infrared Spectroscopy. Applications to Chemical Systems—1*, J. R. Ferraro and L. J. Basile, Eds, 1978, Academic Press, New York, p. 61.
4. COLEMAN, M. M. and PAINTER, P. C., *Proc. Fifth European Symposium on Polymer Spectroscopy*, D. O. HUMMEL, Ed., 1979, Verlag Chemie, Weinheim, p. 49.
5. HOLLAND-MORITZ, K., *ibid.*, p. 93.
6. SIESLER, H. W., *ibid*, p. 137.
7. SIESLER, H. W. and HOLLAND-MORITZ, K., *Infrared and Raman Spectroscopy of Polymers*, 1980, Marcel Dekker, New York.
8. *Analytical Applications of FTIR to Molecular and Biological Systems*, 1980, J. R. Durig, Ed., D. Reidel. Dordrecht.
9. COLEMAN, M. M. and PAINTER, P. C., *J. Macromol. Sci., Rev. Macromol. Chem.*, 1977–1978, **C16**, 197.
10. GRIFFITHS, P. R., *Chemical Infrared Fourier Transform Spectroscopy*, 1975, Wiley and Sons, New York.
11. Nicolet Analytical Instruments Co. Technical notice.
12. COOLEY, J. W. and TUKEY, J. W., *Math. Comput.*, 1965, **19**, 297.
13. KOENIG, J. L., *Analytical Applications of FTIR to Molecular and Biological Systems*, 1979, J. R. Durig, Ed., D. Reidel, Dordrecht, p. 80.
14. ANTOON, M. K., D'ESPOSITO, L., and KOENIG, J. L., *Appl. Spectrosc.*, 1979, **33**, 351.
15. HUGUES, Z. Z. and EL AWADY, A. A., *J. Phys. Chem.*, 1971, **75**, 2954.
16. KOENIG, J. L., D'ESPOSITO, L. and ANTOON, M. K., *Appl. Spectrosc.*, 1977, **31**, 292.
17. KOENIG, J. L. and KORMOS, D., *Appl. Spectrosc.*, 1979, **33**, 349.
18. GOLD, H. S., RECHSTEINER, C. E. and BUCK, R. P., *Anal. Chem.*, 1976, **48**, 1540.
19. ANTOON, M. K., KOENIG, J. H. and KOENIG, J. L., *Appl. Spectrosc.*, 1977, **31**, 518.
20. STAMBOUGH, B., LANDO, J. B. and KOENIG, J. L., *J. Polym. Sci., Polym. Phys. Ed.*, 1979, **17**, 1063.
21. STACH, W. and HOLLAND-MORITZ, K., *J. Mol. Struct.*, 1980, **60**, 49.
22. MOORE, R. S., O'LOANE, J. K. and SHEARER, J. C., *Polym. Preprints*, 1980, **21**, 4.

128   B. JASSE

23. BACHMANN, M. A., GORDON, W. L., KOENIG, J. L. and LANDO, J. B., *J. Appl. Phys.*, 1979, **50**, 6106.
24. PETCAVICH, R. J. and COLEMAN, M. M., *J. Polym. Sci.*, *Polym. Phys. Ed.*, 1980, **18**, 2097.
25. PAINTER, P. C., KESSLER, R. E. and SNYDER, R. W., *J. Polym. Sci.*, *Polym. Phys. Ed.*, 1980, **18**, 723.
26. PAINTER, P. C. and KOENIG, J. L., *J. Polym. Sci.*, *Polym. Phys. Ed.*, 1977, **15**, 1885.
27. O'REILLY, J. M. and MOSHER, R. A., *J. Appl. Phys.*, 1980, **51**, 5137.
28. RENEKER, D. H., MAZUR, J., COLSON, J. P. and SNYDER, R. G., *J. Appl. Phys.*, 1980, **51**, 5080.
29. SENICH, G. A. and MACKNIGHT, W. J., *Macromolecules*, 1980, **13**, 106.
30. COLEMAN, M. M., ZARIAN, J., VARNELL, D. F. and PAINTER, P. C., *Polym. Lett.*, 1977, **15**, 745.
31. COLEMAN, M. M. and ZARIAN, J., *J. Polym. Sci.*, *Polym. Phys. Ed.*, 1979, **17**, 837.
32. COLEMAN, M. M. and VARNELL, D. F., *J. Polym. Sci.*, *Polym. Phys. Ed.*, 1980, **18**, 1403.
33. WELLINGHOFF, S. T., KOENIG, J. L. and BAER, E., *J. Polym. Sci. Polym. Phys. Ed.*, 1977, **15**, 1913.
34. SHELTON, J. R. PECSOK, R. L. and KOENIG, J. L., *Am. Chem. Soc. Sym. Ser.*, USA, 1979, **95**, 75.
35. LIN, S. C., BULKIN, B. J. and PEARCE, E. M., *J. Polym. Sci.*, *Polym. Phys. Ed.*, 1979, **17**, 3121.
36. VANNI, H. and RABOLT, J. F., *J. Polym. Sci.*, *Polym. Phys. Ed.*, 1980, **18**, 587.
37. ISHIDA, H. and KOENIG, J. L., *J. Colloid Interface Sci.*, 1978, **64**, 555; *ibid* 1978, **64**, 565; *J. Polym. Sci.*, *Polym. Phys. Ed.*, 1979, **17**, 615.
38. ISHIDA, H. and KOENIG, J. L., *J. Polym. Sci.*, *Polym. Phys. Ed.*, 1980, **18**, 1931.
39. WOOL, R. P. and STATTON, W. O., in: *Applications of Polymer Spectroscopy*, 1978, E. G. Brame, Ed., Academic Press, New York, p. 185.
40. SIKKA, S. S. and KAUSH, H. H., *Colloid and Polym. Sci.*, 1979, **257**, 1060.
41. SIKKA, S. S., *Polym. Bull.*, 1980, **3**, 123.
42. LEVY, R. L. and WOOL, R. P., *Polym. Preprints*, 1980, **21**, 239.
43. TSE, M. K. and PAIK SUNG, C. S., *Org. Coat. Plast. Chem.*, 1980, **42**, 734.
44. MATSUI, T. and TANAKA, S., *Kobunshi Ronbunshu*, 1980, **37**, 179.
45. SIESLER, H. W., *J. Mol. Struct.*, 1980, **59**, 15.
46. HOLLAND-MORITZ, K., STACH, W. and HOLLAND-MORITZ, I., *J. Mol. Struct.*, 1980, **60**, 1.
47. HOLLAND-MORITZ, K., STACH, W. and HOLLAND-MORITZ, I., *Progr. Colloid and Polym. Sci.*, 1980, **67**, 161.
48. HOLLAND-MORITZ, K. and VAN WERDEN, K., *Makromol. Chem.*, 1981, **182**, 651.
49. HOLLAND-MORITZ, K., HOLLAND-MORITZ, I. and VAN WERDEN, K., *Colloid and Polym. Sci.*, 1981, **259**, 156.
50. SIESLER, H. W., *Polym. Lett.*, 1979, **17**, 453.
51. SIESLER, H. W., *Makromol. Chem.*, 1979, **180**, 2261.
52. BAYER, G., HOFFMANN, W. and SIESLER, H. W., *Polymer*, 1980, **21**, 235.
53. BAYER, G. and WERNER, T., mentioned by SIESLER, H. W., *J. Mol. Struct.*, 1980, **59**, 15.

54. ROSENCWAIG, A. and GERSHO, A., *J. Appl. Phys.*, 1976, **47**, 64.
55. MACDONALD, F. A. and WETSEL, G. C., *J. Appl. Phys.*, 1978, **49**, 2313.
56. ROSENCWAIG, A., *J. Appl. Phys.*, 1978, **49**, 2905.
57. ROCKLEY, M. G., *Appl. Spectrosc.*, 1980, **34**, 405.
58. ROCKLEY, M. G., *Chem. Phys. Lett.*, 1979, **68**, 455.
59. VIDRINE, D. W., *Appl. Spectrosc.*, 1980, **34**, 314.
60. ROYCE, B. S. H., ENNS, J. and TENG, Y. C., *Bull. Am. Phys. Soc.*, 1980, **25**, 408.
61. LAUFER, G., HUNEKE, J. T., ROYCE, B. S. H. and TENG, Y. C., *Appl. Phys. Lett.*, 1980, **37**, 517.
62. ROYCE, B. S. H., TENG, Y. C. and ENNS, J., *Ultrasonics Symp.*, Boston, 1980, p. 652.

Chapter 4

DYNAMICS OF POLYMERS BY
NEUTRON SCATTERING

J. S. Higgins

*Department of Chemical Engineering and Chemical Technology,
Imperial College, London, UK*

SUMMARY

Using the latest high resolution spectrometers, neutron scattering explores molecular motion at frequencies between 10^7 and 10^{14} Hz. This includes vibrations and rotations of side groups, phonon modes in crystals and high frequency conformational changes in solutions, melts and networks. The random variation in scattering power from nuclide to nuclide allows labelling techniques to be used to advantage, while the large neutron mass leads to a relatively short wavelength at low energies (compared to electromagnetic radiation) and gives spatial information simultaneously with the frequency information.

The basic theory of neutron scattering is developed and then applied to calculations for models of polymeric motion. The specific techniques of use in observing polymeric motion are briefly described. A number of experimental examples are discussed where neutron spectroscopy has provided new information on motion of side groups or of the chain backbone.

1. INTRODUCTION

Motion of polymer molecules occurs over many decades of frequency from the slow reorientation of whole molecules in melts just above their glass transition temperatures (T_g), which may take weeks or even years, to the torsions and vibration of side groups at around 10^{14} Hz. As will be seen, the neutron technique[1-4] samples only the upper end of this range with a limit which has currently been pushed down by the ultra-high resolution spin-echo spectrometer to around 10^7 Hz. Within this range fall the molecular vibrations of side chains, the rotations of such side groups, the travelling waves in crystalline samples and the higher frequency backbone motions of polymers in solution and in melts at temperatures well above T_g.

There is considerable overlap in the information obtained with techniques such as NMR relaxation measurements,[5, 6] IR and Raman spectroscopy,[6] high frequency viscoelastic[7] and ultrasonic measurements.[8] However, the unique feature of the neutron scattering technique is that the relatively heavy neutron has large momentum at modest energies. The wave vector changes associated with energy transfer are important and neutron experiments explore regions of momentum transfer (Q) space away from the $Q \approx 0$ region of IR and Raman spectroscopy. Neutron scattering spectra inherently convey direct information about spatial correlations as well as energy fluctuations in a way not available from the other techniques mentioned. In addition, the neutron interacts directly with the scattering nuclei (no optical selection rules) and this interaction varies from isotope to isotope (in particular, there is a large difference between hydrogen and deuterium) thus allowing labelling techniques to be used.

Despite these compelling advantages, neutron scattering has not been extensively applied to the study of polymer dynamics. This is partly because the highest resolution spectrometers have only recently been available and partly because the necessity of taking the sample to the neutron source has meant spectrometers have appeared relatively inaccessible to polymer scientists.

The types of motion mentioned above cover a fairly broad range and, as can be imagined, demand rather different spectrometers for their observation. Four typical spectrometers which cover the required range between them are described in Section 3. A brief summary of the basic theory of the technique is given in Section 2, together with the models of polymer motion commonly used to interpret the neutron data. Sections 4 and 5 then attempt to survey the field of applications by discussing examples of particular problems where neutron scattering techniques have proved to be useful.

2. BASIC THEORY AND MODELS

2.1. Neutron Scattering Principles and the Scattering Laws

Neutron energies in these experiments are very much lower than the nuclear binding energy so the scattering from an isolated nucleus is isotropic and the interaction characterised by a single parameter—the scattering length, or amplitude, b. The scattering cross-section, σ, is the ratio of scattered neutrons to the incident flux, and for an isolated stationary nucleus

$$\sigma = 4\pi b^2 \tag{1}$$

In a scattering sample containing nuclei with spin or for an isotopically impure sample, the scattering amplitude which varies with the spin state of nuclei and from isotope to isotope, thus varies from site to site.

In a scattering event from stationary nuclei there can be no energy transfer, but the wave vector $\mathbf{k}$ of the neutrons changes ($\mathbf{k}$ is a vector in the direction of travel of magnitude $|\mathbf{k}| = 2\pi/\lambda = mv/\hbar$). The momentum transfer $\hbar\mathbf{Q}$ is defined in terms of the change in wave vector on scattering

$$\mathbf{Q} = \mathbf{k}_i - \mathbf{k}_f \tag{2}$$

where the subscripts i and f refer to the initial and scattered beams respectively. However, since there is no energy transfer $|\mathbf{k}_i| = |\mathbf{k}_f|$ and simple trigonometry shows

$$Q = |\mathbf{Q}| = 4\pi/\lambda \sin \theta/2 \tag{3}$$

where θ is the angle of scatter.

The differential cross-section per atom with respect to solid angle Ω contains information about the spatial arrangements of the scattering nuclei. In a sample containing N atoms

$$\frac{d\sigma}{d\Omega} = \frac{1}{N} \left| \sum_n b_n \exp i(\mathbf{Q}.\mathbf{R}_n) \right|^2 \tag{4}$$

where $\mathbf{R}_n$ is the position vector of the nth nucleus. This expression can be manipulated[1,2] to give

$$\frac{d\sigma}{d\Omega} = \{\langle b^2 \rangle - \langle b \rangle^2\} + \langle b \rangle^2 \left| \sum_n \exp i(\mathbf{Q}.\mathbf{R}_n) \right|^2 \tag{5}$$

where $\langle \ \rangle$ implies averaging over all sites.

Only the last term in eqn (5) contains spatial information—the coherent scattering. The coherent cross-section is given by the mean square scattering amplitude, $\sigma_{coh} = 4\pi\langle b\rangle^2$. The first term is constant in a static experiment and forms the incoherent background. The term $\sigma_{inc} = 4\pi(\langle b^2\rangle - \langle b\rangle^2)$ can also be written as $4\pi\langle(b - \langle b\rangle)^2\rangle$, i.e. the mean square deviation of the scattering lengths from their average value. For isotopically pure samples of nuclei such as ^{12}C, ^{16}O which have zero spin, σ_{inc} itself is zero. For hydrogen it is very large. Table 1 lists values of b, σ_{inc} and σ_{coh}, together with the absorption cross-section σ_{abs} for nuclei commonly found in synthetic polymers and their solvents.

TABLE 1

SCATTERING LENGTHS AND CROSS-SECTIONS FOR SOME COMMON ISOTOPES

	$b \times 10^{12}$ cm	$\sigma_{coh} \times 10^{24}$ cm^2	$\sigma_{inc} \times 10^{24}$ cm^2	$\sigma_{abs} \times 10^{24}$ cm^2 (at $1\cdot08$Å)
^{1}H	$-0\cdot374$	$1\cdot76$	80	0.19
^{2}D	$0\cdot667$	$5\cdot59$	2	$0\cdot0005$
^{12}C	$0\cdot665$	$5\cdot56$	0	$0\cdot003$
^{14}N	$0\cdot94$	$11\cdot1$	$0\cdot3$	$1\cdot1$
^{16}O	$0\cdot58$	$4\cdot23$	0	$0\cdot0001$
^{19}F	$0\cdot56$	$3\cdot94$	$0\cdot06$	$0\cdot006$
$^{ave\ 28\cdot06}Si$	$0\cdot42$	$2\cdot22$	0	$0\cdot06$
^{32}S	$0\cdot28$	$0\cdot99$	0	$0\cdot28$
$^{ave\ 35\cdot5}Cl$	$0\cdot96$	$11\cdot58$	$3\cdot5$	$19\cdot5$

In general, if the scattering nuclei are in motion on the time scale of the experiment, $\mathbf{k}_i \neq \mathbf{k}_f$. It is then the double differential cross-section $d^2\sigma/d\Omega\ dE_f$ which is measured in the experiment, and the incoherent term is no longer a flat background, since it carries spatially uncorrelated information about motion of the scattering nuclei. Although for convenience the coherent and incoherent terms are expressed and discussed separately, experimentally they may be very difficult to separate. The scattering cross-sections are related to two space–time correlation functions introduced by Van Hove,[1, 9] $G(\mathbf{R}, t)$ and $G_s(\mathbf{R}, t)$. $G(\mathbf{R}, t)$ expresses the probability that if there is a nucleus at position $\mathbf{R}_i$ at time $t = 0$, there will be another nucleus at position $\mathbf{R}_j$ at time t. $G_s(\mathbf{R}, t)$ is the self-correlation function, expressing correlations between the same nucleus at time 0 and t.

$$\frac{d^2\sigma_{coh}}{d\Omega\ dE_f} = \frac{k_f}{k_i}\frac{\langle b\rangle^2}{2\pi\hbar}\iint d\mathbf{R}\ dt\ \exp\{i(\mathbf{Q} \cdot \mathbf{R} - \omega t)\}G(\mathbf{R}, t) \qquad (6)$$

$$\frac{d^2\sigma_{inc}}{d\Omega\,dE_f} = \frac{k_f}{k_i}\frac{(\langle b^2\rangle - \langle b\rangle^2)}{2\pi\hbar}\int\int d\mathbf{R}\,dt\,\exp\{i(\mathbf{Q}\cdot\mathbf{R}-\omega t)\}\,G_s(\mathbf{R},t)\quad(7)$$

where $\qquad\qquad \hbar\omega = \Delta E = E_f - E_i = \dfrac{\hbar^2}{2m}(k_f^2 - k_i^2)\qquad\qquad(8)$

The integrals in eqns (6) and (7) contain all the information about the scattering system and are usually called the scattering laws with

$$S(\mathbf{Q},\omega) = \frac{1}{2\pi}\int\int \exp\{i(\mathbf{Q}\cdot\mathbf{R}-\omega t)\}\,G(\mathbf{R},t)\,d\mathbf{R}\,dt\quad(9)$$

and an analogous definition for $S_s(\mathbf{Q},\omega)$ in terms of $G_s(\mathbf{R},t)$. If only the spatial Fourier transforms (in eqn (9) and its incoherent analogue) are performed, the quantities obtained are the so-called intermediate scattering functions $S(\mathbf{Q},t)$ and $S_s(\mathbf{Q},t)$.

2.2. Model Scattering Functions for Moving Molecules
Although, in principle, the experimentally observed cross-sections can be Fourier transformed to obtain the correlation functions, in practice the $\mathbf{Q}$ and ω ranges are usually limited so it is more usual to calculate scattering laws based on models of the molecular motion and compare these with the data. The derivations of those functions are not usually difficult but an extended discussion would be out of place in the present context. The final expressions will simply be quoted and the important features to be expected in the experimental data will be discussed.

2.2.1. Vibrations
In the classical limit for a molecular vibration of frequency ω_0 and amplitude $\mathbf{u}_0$, which gains or loses energy in a neutron scattering event,

$$\frac{d^2\sigma}{d\Omega\,dE_f} = \frac{k_f}{k_i}\sum_{n,m}\langle b_n b_m\rangle\,\exp[i\mathbf{Q}\cdot(\mathbf{n}-\mathbf{m})]\,\exp(-2W)\times$$

$$[\delta(\hbar\omega) + \tfrac{1}{2}(\mathbf{Q}\cdot\mathbf{u}_0)^2\,\delta(\hbar\omega - \hbar\omega_0) + \tfrac{1}{2}(\mathbf{Q}\cdot\mathbf{u}_0)^2\,\delta(\hbar\omega + \hbar\omega_0) + \ldots]\quad(10)$$

where $\mathbf{n}$ is the position vector of the nth nucleus and $\mathbf{u}_n$ its displacement. This classical limit shows some of the more important features of scattering from molecular vibrations. The δ functions define an elastic component $\hbar\omega = 0$ and two inelastic components $\hbar\omega = \pm\hbar\omega_0$ corresponding to neutrons gaining or losing a quantum of vibrational energy (the higher terms in the series correspond to multi-quantum effects). The intensity of the inelastic peaks is proportional to $(\mathbf{Q} . \mathbf{u}_0)^2$, i.e. it increases with $\mathbf{Q}$ and with the amplitude of the vibration observed. In the simplified form of eqn (10), both the elastic and inelastic components are attenuated by the term $\exp(-2W)$. For real situations the inelastic factor may be more complex and energy dependent. This attenuation arises because interference effects between atoms at different sites are smoothed out by thermal vibrations. In eqn (10) $\exp(-2W) = \exp(-(W_n + W_m))$, where $W_n = \frac{1}{2}\langle (\mathbf{Q} . \mathbf{u}_n)^2 \rangle$. The single quantum inelastic scattering in eqn (10) can be manipulated to show explicitly the coherent and incoherent terms

$$\frac{d^2\sigma^{(1)}}{d\Omega \, dE_f} = \frac{k_f}{k_i} \sum_{n,m} \langle b \rangle^2 \exp(i\mathbf{Q} . (\mathbf{n} - \mathbf{m})) \frac{(\mathbf{Q} . \mathbf{u}_0)^2}{2} (\delta(\hbar\omega - \hbar\omega_0) +$$

$$\delta(\hbar\omega + \hbar\omega_0)) + \frac{k_f}{k_i}(\langle b^2 \rangle - \langle b \rangle^2) \frac{N(\mathbf{Q} . \mathbf{u}_0)^2}{2} (\delta(\hbar\omega - \hbar\omega_0)$$

$$+ \delta(\hbar\omega + \hbar\omega_0)) \tag{11}$$

The first term is the coherent scattering and contains a phase factor relating atoms at $\mathbf{n}$ and $\mathbf{m}$. This imposes conditions on the $\mathbf{Q}$ values at which the scattering can take place—conditions which are absent for the second, incoherent, term.

Note that because of the term $\mathbf{Q} . \mathbf{u}_0$ both coherent and incoherent scattering from oriented samples will show direction properties with intensity at a maximum when $\mathbf{Q}$ is parallel to $\mathbf{u}_0$ and zero when $\mathbf{Q}$ is perpendicular to $\mathbf{u}_0$. In unoriented samples averaging over all molecules removes this condition.

In extended arrays of ordered molecules (or atoms) the intermolecular motions of the weakly coupled molecules constitute sets or branches of normal modes which may be represented by travelling waves or phonons. The frequencies of these modes $\omega(\mathbf{q})$ are characterised by a wave vector $\mathbf{q}$, and the complex motions reflect the symmetry and periodic properties of the lattice. Because the motion in adjacent unit cells is correlated, the spatially sensitive coherent neutron scattering cross-section is especially sensitive. Since both energy and wave vector or momentum are conserved in this scattering process, sharp peaks are observed when either the energy transfer ω or momentum transfer $\mathbf{Q}$ is scanned and the *two* conservation conditions are satisfied

$$\mathbf{Q} = \mathbf{g} + \mathbf{q}$$
$$\omega(\mathbf{q}) = E_f - E_i$$

where **g** is a reciprocal lattice vector (the mode frequencies are periodic with the lattice periodicity).

For N molecules there are $3N - 6$ intermolecular modes possible; considering M molecules per unit cell, these lattice modes are distributed into $6M$ branches (3 translational and 3 librational). Along high symmetry directions in the crystal a clear distinction may be made between modes parallel (longitudinal) and transverse (displacements **u** perpendicular) to the direction of propagation **g**. Analysis of the dispersion of the phonon energy $\omega(\mathbf{q})$ with **q** leads to detailed information on the intermolecular potential.

The incoherent scattering from a crystal contains much less detailed information. In particular, the δ function in **Q** disappears and with it much directional information. All possible energy transfers will be observed at any given value of **Q**. Assuming one atom in a cubic unit cell, the incoherent term in eqn (10) becomes

$$\frac{\mathrm{d}^2\sigma^{(1)}}{\mathrm{d}\Omega\,\mathrm{d}E_f} = N\frac{k_f}{k_i}(\langle b^2\rangle - \langle b\rangle^2)\frac{Q^2\langle u^2\rangle}{2M}\frac{\exp(-2W)}{\sinh(\hbar\omega/kT)}\frac{Z(\omega)}{\omega} \quad (12)$$

where $Z(\omega)$ is the density of phonon states:

$$Z(\omega) = \frac{\hbar\omega}{3N}\sum_{\mathbf{q}j}\frac{\delta(\hbar\omega \mp \hbar\omega_j(\mathbf{q}))}{\omega_j(\mathbf{q})} \quad (13)$$

$\langle u^2\rangle$ is the mean square amplitude of vibration and M the mass of the vibrating atom. The sinh term in the denominator arises from the Bose–Einstein population factor for the energy levels.

The scattering law $S_s(\mathbf{Q}, \omega)$ is obtained from $\mathrm{d}^2\sigma/\mathrm{d}\Omega\,\mathrm{d}E_f$ by

$$S_s(\mathbf{Q}, \omega) = \frac{k_i}{k_f}\frac{\hbar}{N(\langle b^2\rangle - \langle b\rangle^2)}\frac{\mathrm{d}^2\sigma}{\mathrm{d}\Omega\,\mathrm{d}E_f} \quad (14)$$

For comparison with optical spectra obtained at zero Q the experimental scattering laws are sometimes extrapolated to obtain the zero Q amplitude weighted density of states[4, 10] $g(\omega)$ (rather than $Z(\omega) \approx g(\omega)/\langle u^2\rangle$).

$$g(\omega) = \omega\sinh\frac{\hbar\omega}{2kT}\underset{Q\to 0}{\mathrm{Lim}}\left[\frac{S_s(\mathbf{Q}, \omega)}{Q^2}\right] \quad (15)$$

The thermal effects of the population factor have been divided out and the Q variation in the $Q^2\langle u^2\rangle$ term and the Debye–Walker factor removed by the extrapolation to zero Q.

2.2.2. Rotational Motion of Side Chains

The continuous rotation of molecules or molecular side groups can essentially be considered to be in the classical (non-quantised) limit for the time scale of the neutron experiments. It therefore appears in the neutron spectrum as a Doppler broadening of the elastic (and incidentally, the inelastic) component in eqn (10). In particular, rotation about a fixed centre of mass (as, for example, a side chain in a polymer sample below T_g) gives a two component scattering in the near zero energy transfer region with an elastic component and a quasielastic component.[11-13]

$$S_s(\mathbf{Q}, \omega) = A_0(Q)\delta(\omega) + \frac{1}{\pi}(1 - A_0(Q))F(\omega) \tag{16}$$

The elastic incoherent structure factor $A_0(Q)$ which governs the relative intensities of the two components is, in turn, governed by the shape swept out by the rotating nuclei. For a hydrogen nucleus in a methyl group rotating about a mixed axis between three equivalent sites,[11]

$$A_0(Q) = \frac{1}{3}(1 + 2j_0(Qr)) \tag{17}$$

where r is the radius of gyration, and $j_0(x)$ is a zero order spherical Bessel function.

The broadened component is Lorentzian in form

$$F(\omega) = \frac{2\tau_r/3}{1 + \omega^2(2\tau_r/3)^2} \tag{18}$$

with a full width at half maximum $4\tau_r/3$, where τ_r is the time between jumps. For more complex motion, various forms for $A_0(Q)$ and $F(\omega)$ have been derived. The general principle of a broadened component governed by the rotational frequency is maintained. $A_0(Q)$ is essentially the spatial Fourier transform of the shape swept out around their fixed centre of gravity by the scattering nuclei. A large rotational volume Fourier transforms to give an $A_0(Q)$ peaked close to zero Q and in the limit as the volume tends to infinity (i.e. we have no fixed centre of mass and the sample becomes liquid) the elastic component disappears at finite Q.

2.2.3. Main Chain Motion in Solution and in the Melt

In liquids the elastic component in eqns (10) and (16) are completely replaced by Doppler broadened quasielastic peaks. Although a polymer solution or melt is a true liquid in the sense that there is no fixed centre of mass, the experimental Q-value defines whether the motion of all or only part of the molecule is observed. The exponents $(\mathbf{Q} \cdot \mathbf{R})$ in eqns (5), (6) and (7) ensure that the scattering from self- or pair-correlations over a distance R will be maximum when $Q = 1/R$ (or multiples of $1/R$ for a periodic array such as a crystal). The Q-value will thus 'pick out' different scales of molecular motion. For $Q < 1/R_g$ where R_g is the radius of gyration of the polymer molecule, overall centre of mass diffusion is dominant. If this is (as is usual) Fickian diffusion, both the scattering law—$S(\mathbf{Q}, \omega)$—and the intermediate scattering laws $S(\mathbf{Q}, t)$ have a particularly simple dependence on $\mathbf{Q}$ and the diffusion coefficient, D, for both coherent and incoherent scattering,

$$\frac{S(\mathbf{Q}, \omega)}{S(\mathbf{Q})} = \frac{\Delta\omega}{\Delta\omega^2 + \omega^2} \tag{19}$$

$$\frac{S(\mathbf{Q}, t)}{S(\mathbf{Q})} = \exp(-\theta) \tag{20}$$

where θ is the normalised time (often written Γt—see below).

Parameters extracted from the data are the half width at half maximum of $S(\mathbf{Q}, \omega)$, called $\Delta\omega$, and the slope of a plot of $\ln S(\mathbf{Q}, t)$ against t, called the inverse correlation time θ/t. For a simple exponential decay such as that in eqn (20), θ/t is the time taken for the signal to fall to $1/e$ of its value at time $t = 0$. $\Delta\omega \equiv \theta/t \equiv DQ^2$ for diffusional motion but this range is only accessible with neutrons for the smallest molecules. Centre of mass diffusion is conventionally observed for polymer solutions using photon correlation spectroscopy.[14] For larger values of Q, the internal motion of the molecule is observed as it constantly changes its conformation. This continually changing conformation causes fluctuation in the distances between points on the chain, and the entropy gain as the length of any segment increases from its equilibrium value causes that segment to act as a spring under tension. The model developed by Rouse[15] represents the polymer molecule as a series of such springs of length a connected by beads. The effect of the surrounding solvent molecules is described by a frictional drag on the beads. Zimm[16] modified the model to take account of hydrodynamic coupling between the beads existing in many real situations. Despite the loss of any chemical structure for Rouse chains (the only variable parameter is the spring length a) these models have been extremely successful in describing viscoelastic behaviour of polymer solutions excepting only the highest frequencies where local vibrations and rotations are dominant and the chemical structure can no longer be ignored.

Pecora[17] and de Gennes[18, 19] calculated the intermediate scattering laws for Zimm chains in the region $R_g^{-1} \ll Q \ll a^{-1}$. Both the coherent and the incoherent scattering show an unusual dependence on $\theta^{2/3}$ where θ is the normalised time and

$$\theta/t_{\text{Zimm}} \propto \frac{k_B T}{\eta_s} Q^3 \quad (\eta_s \text{ is the solvent viscosity}) \tag{21}$$

Fourier transformation of the intermediate functions does not lead to a particularly simple analytical form for $S(\mathbf{Q}, \omega)$ and $S_s(\mathbf{Q}, \omega)$ but the width functions $\Delta\omega_{\text{inc}}$ and $\Delta\omega_{\text{coh}}$ show the same Q^3 variation as θ/t.

TABLE 2

NUMERICAL PREFACTORS FOR THE WIDTH FUNCTION CALCULATED FOR SEVERAL MODELS OF MOLECULAR MOTION

Model	Form of Q dependence of width function		Numerical coefficients for	
			Θ/t	$\Delta\omega$
Rouse $R_g^{-1} < Q < a^{-1}$	$\dfrac{k_B T a^2}{\xi_0} Q^4$	coherent	0·33	0·02
		incoherent	0·106	0·03
Zimm $R_g^{-1} < Q < a^{-1}$	$\dfrac{k_B T}{\eta_s} Q^3$	coherent	0·0375[a]	0·0303
		incoherent	0·0923	0·0413
Fickian diffusion	DQ^2	coherent	1	1
		incoherent	1	1

θ solvent conditions (bracketing the Rouse and Zimm $\Delta\omega$ coefficients)

[a] Assuming preaveraged Oscen Jensor.[89]

In circumstances where the hydrodynamic interactions may be screened out (i.e. a polymer melt) the original Rouse description may be more valid. This leads to an even slower $\theta^{1/2}$ variation with normalised time and

$$\theta/t_{\text{Rouse}} \propto \frac{k T a^2}{\xi_0} Q^4 \tag{22}$$

ξ_0 is the friction factor per segment. Table 2 summarises the numerical coefficients in eqns (21) and (22) for θ/t and $\Delta\omega$ for both coherent and incoherent scattering.

For distances shorter than a local vibrational fluctuations are explored. There are no explicit models for this region but both qualitative discussion in terms of Brownian motion and limited numerical calculation[17] suggest a return to the Q^2 dependence of $\Delta\omega$ and θ/t. The Rouse model depends on the Gaussian nature of the conformation of segments within one spring length a. Realistically then, even for the most flexible polymers, a is expected to be of the order 10 Å or more. The lowest Q values obtained with the high resolution quasielastic spectrometers described in Section 3 is around 0.03 Å^{-1}. The experiments inherently explore therefore the transition region $Qa \approx 1$ and the short range local motions. For this reason, the recent development of model calculations by Akcasu et al.[20] has been particularly valuable. They calculated the correlation functions $S(\mathbf{Q}, t)$ continuously over the range from $Q < R_g^{-1}$ to $Q > a^{-1}$ for the Rouse and Zimm models. Clearly at the upper inequality at which motion of a single 'bead' dominates, the model becomes unrealistic but as analysis in Section

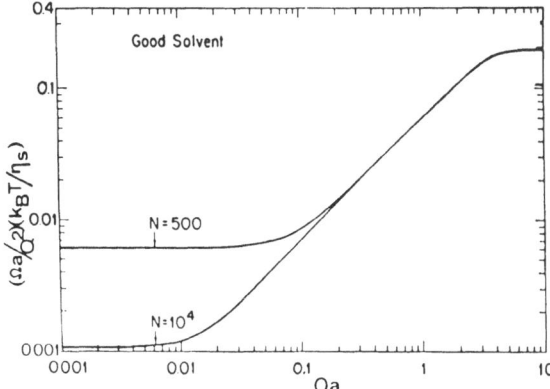

FIG. 1.   Calculated values of $[\Omega a/Q^2][k_B T/\eta_s]$ vs Qa for a good solvent and for two lengths of polymer chain Na.[20]

5 shows, this theory offers the first objective method of assessing the neutron scattering results. The parameter extracted from fitting the correlation functions (or intermediate scattering functions) is the first cumulant (or initial slope) of $\ln (S(\mathbf{Q}, t))$

$$\Omega = \lim_{t \to 0} \frac{d}{dt} \ln S(\mathbf{Q}, t) \qquad (23)$$

For Fickian diffusion $\Omega \equiv \theta/t$, for the range $R_g^{-1} < Q < a^{-1}$, $\Omega \equiv \sqrt{2}\theta/t$.

Figure 1 shows the variation of Ω with Qa over the whole range of motion described for the Zimm model of a polymer in a good solvent. Since T and η_s are fixed, there is only one variable parameter, the spring length a. Akcasu et al.[20-22] were able to develop the discussion of Ω in some detail, looking at the effect of solvent interactions, solution concentrations and the strength of hydrodynamic interactions. As will be seen in Section 5, this approach now forms the basis for interpretation of the neutron results.

The major difference when extending the previous discussion to polymer melts lies in the entangled nature of the chains. However, although there has been some discussion of the possible nature of the scattering from polymers moving against entanglements,[23] it has been shown[24, 25] that such effects only become important when motion is observed over distances greater than 30 Å. The neutron experiments therefore observe on a scale where motion of single chains dominates and the models described above are adequate for interpreting the data to a first approximation.

3. EXPERIMENTAL TECHNIQUES

3.1. General Introduction

In order to determine the inelastic neutron scattering cross-sections discussed in the previous section, it is normally necessary to define the incident energy more or less precisely using some form of monochromator and to then analyse the energy of the scattered neutrons as a function of scattering angle. To date, most neutron experiments have used the thermal neutrons from reactors with a Maxwellian wavelength distribution peaked at 1·4 Å (in some cases shifted to longer wavelengths by the use of a cold source, as described below).

A slice of wavelengths is taken out of this distribution using either mechanical choppers (time-of-flight and spin-echo spectrometers use mechanical monochromation) or using Bragg reflection from a suitably oriented crystal (the triple axis and back-scattering spectrometers use this method). Analysis of the scattered neutrons can again use Bragg reflection from a suitable crystal, or measurement of the time taken to cover a measured flight path (time-of-flight analysis) of a pulsed beam.

Recently, a new generation of neutron sources[26-28] is being constructed where accelerated beams of charged particles (electrons or protons) are incident on a target of high atomic number and produce intense bursts of high energy neutrons. The high energies and short wavelengths of these neutrons make them unique for investigation of large energy transfers and high wave vectors. This area is not typically associated with polymer dynamics; however, the neutrons can be moderated to wavelengths of the same order as reactors and then it is the extremely high intensity in the pulse which is exploited. Generally, spectrometers on pulsed sources[28] use the pulsed nature of the beam together with time-of-flight analysis to obtain good energy resolution. Many such spectrometers are at the design stage at present[27] but none has yet been constructed for the low energy (high resolution) range of interest for studying polymer dynamics. For this reason, the spectrometers described here are all the most recently constructed reactor instruments of their type. They are all at the high flux reactor at the Institut Laue–Langevin[29] in Grenoble which is jointly owned by the scientific communities of France, Germany and the UK. All four instruments are at the end of neutron guides directed at a cold source. The guides conduct the neutrons to areas away from the reactor where there is more space and lower background 'noise'. The cold source is a container of liquid deuterium close to the reactor core where the neutrons are scattered many times and lose energy to finish with a wavelength distribution shifted to long wavelengths (typically 5–10 Å).

3.2. The Triple-Axis Spectrometer

Figure 2 shows the layout of the triple-axis spectrometer IN12.[30] The spectrometer uses neutrons from a guide on a cold source with a useful flux of wavelengths between 2·3 and 6·3 Å. Apart from the wavelength range it is typical of all such spectrometers.[31, 32] The whole spectrometer rotates about the monochromator (M) which, on IN12, is normally pyrolitic graphite (spectrometers for shorter incident wavelengths use copper or germanium monochromator crystals). The analyser and detector arms rotate about the sample (S) and the detector (D) about the analyser crystal (A). The analyser can often be the same material as the monochromator but could also be chosen from any crystal suitable for covering the desired energy transfer range. The three axes of the apparatus are thus M, S and A. In between are three collimators C_{MS}, C_{SA} and C_{AD}. These are well defined narrow slots in the beam direction covered in neutron absorbing material (cadmium or gadolinium) which reduce beam divergence. The three-axis construction, which allows variation of E_i, E_f and θ, gives almost unlimited flexibility in choosing $\mathbf{Q} = \mathbf{k}_i - \mathbf{k}_f$ and $\Delta E = E_i - E_f$ independently. It is thus possible to map the whole of $\mathbf{Q}$–E space and the spectrometer is particularly suited to the tracing of phonon dispersion curves. Since the first one designed by Brockhouse,[33] these spectrometers have changed little and have been one of the basic tools for neutron scattering use in solid state physics. Since each point is counted separately, however, this is not a particularly fast spectrometer for covering large areas of $\mathbf{Q}$–E space. Such experiments are better tackled using the time-of-flight technique. The

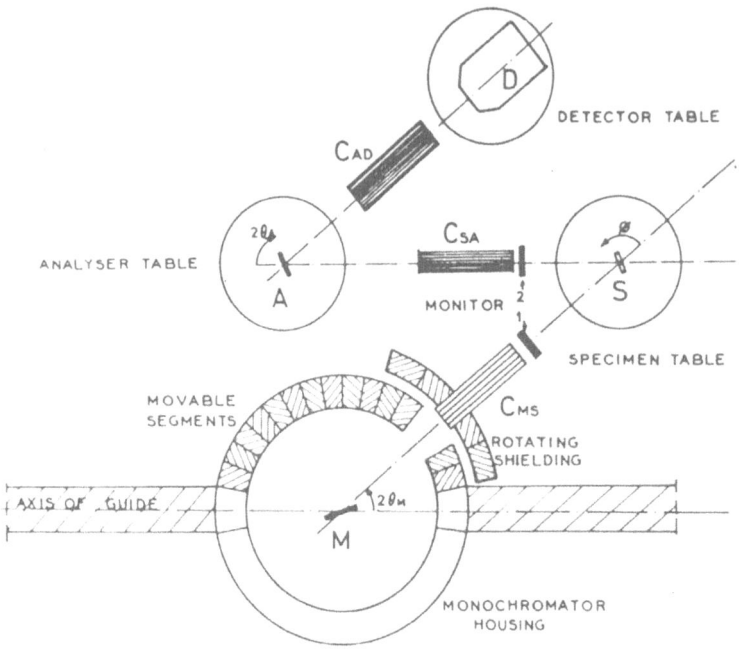

FIG. 2.   The IN12 triple-axis spectrometer at the Institut Laue–Langevin. M,
monochromator; A, analyser; S, sample; C, collimator; D, detector.

spectrometer measures the differential cross-section $(d^2\sigma)/(d\Omega\,dE_f)$
directly and IN12 has a best resolution of $\Delta E/E$ around 0·5%.

3.3. The Time-of-Flight Spectrometer

Time of flight spectrometers[34] increase the counting rate by using banks of
detectors at fixed scattering angles but lose the flexibility of uncoupled ΔE
and **Q**. There is more variation between spectrometers possible here, but
in general a pulsed monochromatic beam produced either by rotating a
crystal monochromator or by phased mechanical choppers is scattered by
the sample and the neutron flight time to the detectors is measured. If the
flight path is d, then time of flight, $\tau = t/d = \text{velocity}^{-1}$. Hence ΔE is
known but in each experiment E_i is constant, so that Q is coupled to ΔE

since

$$Q = \sqrt{k_i^2 + k_f^2 - 2k_i k_f \cos\theta}$$

and

$$\Delta E = \left(\frac{\hbar^2 k_i^2}{2m} - \frac{\hbar^2 k_f^2}{2m}\right) \quad (24)$$

In quasielastic experiments, ΔE is very small so this coupling to Q may not be important. Similarly, in non-crystalline samples, dispersion of molecular modes is often small so that inelastic neutron scattering spectra can be used to give information on these modes without requiring the flexibility of precise choice of ΔE and $\mathbf{Q}$ given by the triple axis apparatus. In these circumstances, the enormous increase in count rate given by using

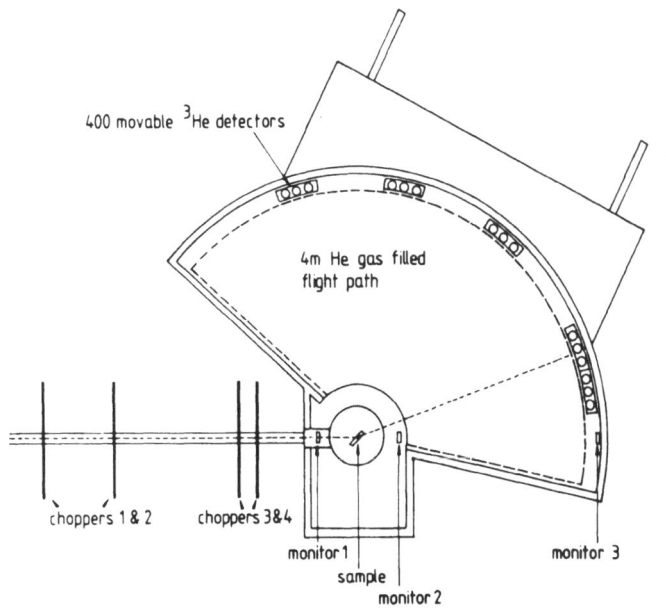

FIG. 3.   The time-of-flight spectrometer IN5 at ILL.

large detector banks has made the time-of-flight spectrometer the work-horse in neutron investigation of dynamics of liquids and amorphous materials.

Figure 3 gives the layout of the IN5 spectrometer[35] at ILL. This uses phased choppers, two to select a wavelength, one to remove harmonics and one to reduce the pulse repetition rate where necessary, to avoid frame overlap. Neutrons scattered by the sample then travel through a helium-filled four-metre flight path to banks of ^{3}He detectors covering scattering angles from 1° to 140°. Monitors in the direct beam at measured distances apart allow precise determination of E_i. The number of neutrons arriving in each detector as a function of time gives directly the double differential cross-section as a function of time of flight τ. This must be related then to the more usual energy cross-section by

$$\frac{\partial^2 \sigma}{\partial \Omega \, \partial E_f} = \frac{-\tau^3}{m} \frac{\partial^2 \sigma}{\partial \Omega \, \partial \tau} \tag{25}$$

Around the elastic peak the τ^3 term does not vary very fast, but across the inelastic spectrum this correction has an important effect on relative intensities of different parts of the spectrum. It is interesting to note that extraction of $S(\mathbf{Q}, \omega)$ from $\partial^2 \sigma / \partial \Omega \, \partial E_f$ (see eqns (6–9)) introduces a factor k/k_0 which is numerically equal to τ/τ_0 so that $S(\mathbf{Q}, \omega)$ is effectively multiplied by τ^4 in the time-of-flight spectrum. However, the quantity required from these inelastic spectra is often the density of states (see eqns (12–16)) in which $S(\mathbf{Q},\omega)$ has to be multiplied by $\omega \sinh(\hbar\omega/kT)$. This factor varies approximately as ω^2 over a limited energy range. Now $\hbar\omega = \Delta E$ which will vary over a limited range roughly as E_f^2—i.e. as $1/\tau^4$—so the two factors cancel and the time-of-flight spectrum is often very close in shape to the density of states finally required. The energy resolution of such spectrometers depends both on the rotation rate of the choppers and the value of E_i chosen, but the best value of $\Delta E/E$ for IN5 is 1%. In terms of quasielastic scattering, this leads to a best resolution for 10 Å incident neutrons of around 20 μeV ($\omega \approx 3 \times 10^{10}$ Hz).

3.4. The Back-Scattering Spectrometer

The 10^{10} Hz resolution limit of IN5 represents the best that is likely to be achieved with the time-of-flight technique, mainly for reasons of flux limitation. It is normally possible to detect motion about one order of magnitude in energy less than the resolution of the spectrometer—thus IN5 allows detection down to $\sim 10^9$ Hz. However, the interesting range for polymer backbone motion has been even lower in energy. In order to achieve better resolution the back-scattering spectrometer sacrifices the wide window in energy allowed on other spectrometers and concentrates the available flux in a narrow range observed with very high resolution.

The back-scattering technique uses a crystal monochromator set at a 180° Bragg angle, which gives an extremely sharp energy spectrum (~ 1 μeV or 10^9 Hz). Its basic construction is that of a triple axis spectrometer in which a mechanical drive Doppler shifts the neutron incident energies and scans small energy changes on scattering from the sample. The IN10 spectrometer[36] at ILL is shown schematically in Fig. 4. Neutrons arrive in a neutron guide from the reactor and are incident on a monochromator crystal. This is silicon, oriented so that the (111) planes Bragg reflect 6·2 Å neutrons at 180°. Because of this 180° reflection angle, θ, the wavelength spread which is proportional to $\cot(\theta/2)\,d\theta$ is very narrow. This 6·2 Å beam travels back along its path to a graphite crystal and is there reflected at 45° through a chopper (the function of which will be explained later) to a sample. Neutrons scattered by the sample are incident on silicon crystal analysers oriented with the (111) planes again in back reflection for 6·2 Å neutrons. The analysers are on curved mountings focused on detectors just behind the sample position. If the sample scatters inelastically, the intensity counted will be diminished since some neutrons no longer reach the analysers with the necessary 6·2 Å wavelength. However, the mono-chromator crystal is mounted on a Doppler drive which imparts a small shift in energy to the 6·2 Å reflected neutrons. Only if they lose or gain to

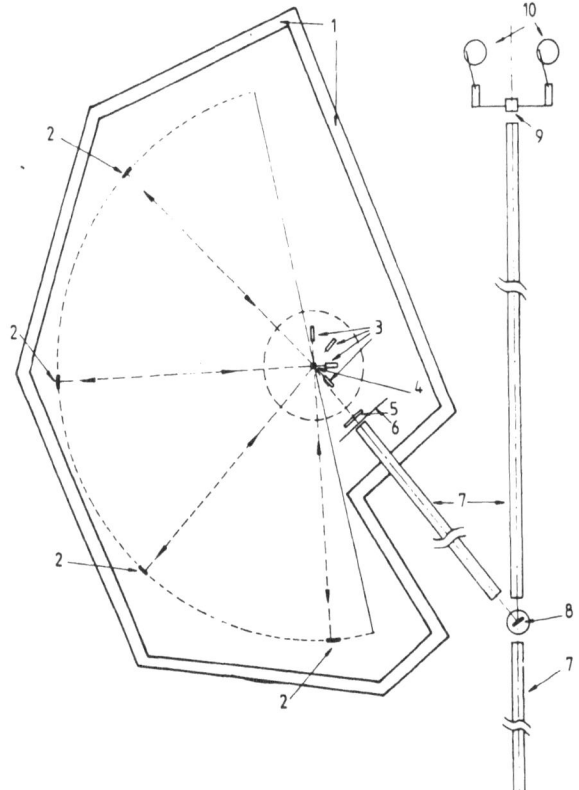

FIG. 4.   The back-scattering spectrometer IN10 at ILL. 1. Shielding; 2. Analyser crystals; 3. ^{3}He detectors; 4. Sample; 5. Monitor; 6. Auxiliary chopper; 7. Neutron guide; 8. Graphite crystal; 9. Monochromator crystal; 10. Doppler drive.

the sample the energy they have gained or lost at the Doppler drive, so that they again have exactly 6·2 Å wavelength, will the neutrons be back-reflected at the analysers and reach the detector.

(The auxiliary chopper coarsely pulses the neutrons and time-of-flight analysis allows discrimination against neutrons scattered directly into the detectors). In this way, a small energy scan ($\pm \sim 12 \mu eV$) can be made with a very sharp (1 μeV) resolution. As can be seen, in this case resolution has been won at the expense of the width of the energy window.

Other limitations concern the Q-range and resolution. In order to increase the counting flux, the focusing analyser crystal banks are made relatively large, relaxing the angular resolution, and giving a $\Delta Q/Q$ of order 10%. These analysers can, of course, be masked to give better $\mathbf{Q}$ resolution but this is only done for special circumstances because of the severe flux penalties. For similar reasons the lowest angles of scatter used are limited and the smallest Q value is 0·07 Å^{-1}. Even so, there is a danger of some interference at the lowest Q values, and $\Delta Q/Q$ becomes very large. Within these limitations the apparatus has been very successfully used in observation of tunnelling splittings of order a few μeV and of quasielastic scattering arising from motion of about 10^9–10^{10} Hz in liquids, macromolecules and liquid crystals.

3.5. The Spin-Echo Spectrometer

This technique originally devised by F. Mezei[37] uses the precession of the neutron spin in a magnetic guide field as a counter to measure very small changes in neutron velocity. In order to explain the method, it is necessary to follow the path of neutrons through the apparatus, which is shown schematically in Fig. 5.

The spectrometer has[38, 39] two identical arms on either side of the sample position, each consisting of a length of solenoid providing a magnetic guide field directed along the flight path, and a $\pi/2$ spin turn coil. The incoming neutrons are roughly monochromated using a velocity selector—the wavelength spread $\Delta\lambda/\lambda$ is about 10%. This monochromator is, as will be seen, unnecessary for the *energy* resolution[39b] but dominates the Q-resolution. After the polariser the neutron spins are aligned along the flight path. The $\pi/2$ spin turn coil rotates the neutron spin direction which starts precessing about the guide field. The precession rate of each neutron is the same—given by the Larmor precession frequency ($= \omega_L = 2\mu_n B_0/\hbar$ where μ_n is the magnetic moment and B_0 the guide field strength). However, the number of precessions will be governed by the length of time taken to traverse the guide field, i.e. the neutron velocity. A beam containing a

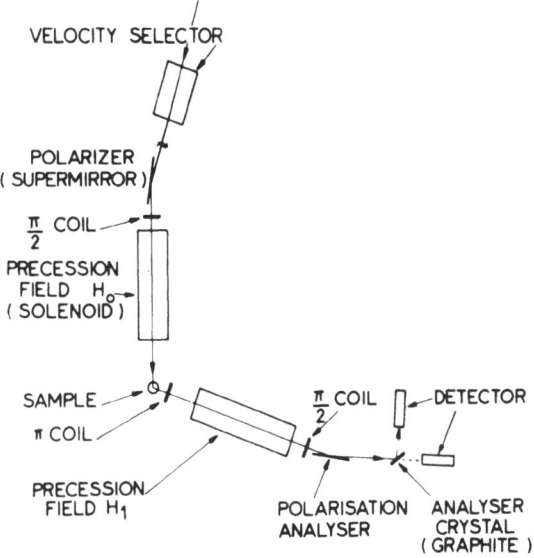

FIG. 5.   The spin-echo spectrometer IN11 at ILL.

wavelength spread loses its initial phase coherence as it travels through the
guide field and in analogy with NMR the spins will have fanned out with
respect to each other. In the π turn coil, located at the sample, the neutrons
make a 180° precession *about a field perpendicular to the guide field* $(H^{\perp})$.
This has the effect of reflecting the distribution in the $H^{\perp}$ axis. Those spins
that were furthest behind in the fan of precession angles are now furthest
ahead and vice versa. Thus, if the length of the second guide field is
adjusted to be identical to the first, all the spins will once more be aligned at
the second $\pi/2$ turn coil and 100% polarisation in the flight path direction
will be observed at the analyser.

To observe quasielastic scattering, the number of precessions in each
guide field is set equal for elastic events so that small changes in energy at
the sample result in a neutron performing unequal precessions in the two
fields and produce a reduction in the polarisation received at the analyser.
The net polarisation $\langle P_z \rangle$ is then given by

$$\langle P_z \rangle = \int_0^{\infty} I(\lambda)\,d\lambda \int_{-\infty}^{+\infty} P(\lambda,\,\delta\lambda) \cos \frac{2\pi N_0 \delta\lambda}{\lambda_0}\, d(\delta\lambda) \qquad (26)$$

where $I(\lambda)\,d\lambda$ is the initial wavelength spread and $P(\lambda,\,\delta\lambda)$ is the probability that a neutron of wavelength λ will be scattered with a wavelength

change $\delta\lambda$. Since $\delta\lambda$ can be transformed into an energy change $\delta E = \hbar\omega$ this is, of course, just the scattering law $S(\mathbf{Q}, \omega)$

$$\int P(\lambda, \delta\lambda)\,d(\delta\lambda) = \int S(\mathbf{Q}, \omega)\,d\omega \qquad (27)$$

N_0 is the number of precessions made by a neutron of the mean wavelength λ_0. Now since

$$E = \frac{\hbar^2}{2m}\frac{1}{\lambda^2} \qquad (28)$$

$$\hbar\omega = \delta E = \frac{\hbar^2}{m}\frac{\delta\lambda}{\lambda^3} \text{ or } \delta\lambda = \frac{m\lambda^3}{2\pi\hbar}\omega$$

and

$$\langle P_Z \rangle = \int_0^\infty I(\lambda)\,d\lambda \int_{-\infty}^\infty S(\mathbf{Q}, \omega) \cos\left\{\left(\frac{N_0 m\lambda^3}{\hbar\lambda_0}\right)\omega\right\}\,d\omega \qquad (29)$$

The expression $N_0 m\lambda^3/\hbar\lambda_0$ has the dimensions of time, and can be designated $t(\lambda)$.

P_Z, then, is just the Fourier transform of $S(\mathbf{Q}, \omega)$, i.e. the time correlation function $S(\mathbf{Q}, t)$ is observed directly in spin-echo experiments. The experimental time scale $t(\lambda)$ is governed through N_0 by the strength of the magnetic guide field. For neutrons of 8 Å the time scale covers about two decades from 10^{-9} to 10^{-7} secs. For study of polymer dynamics, direct observation of $S(\mathbf{Q}, t)$ has considerable advantages as will appear below when analysis of the experimental data is considered.

As with the back-scattering spectrometer, the very high resolution ($\langle 0\cdot 1\ \mu\text{eV}, 10^{-8}$ Hz) is won at the expense of Q resolution. In this case, the angular definition of Q is very good, since the beam has to be highly collimated within the guide fields. $\Delta\lambda/\lambda$ is, however, quite broad ($\sim 10\%$) in order to increase flux and this imposes a 10% limit on $\Delta Q/Q$. On the other hand, the tight collimation does allow use of small ($\sim 1°$) scattering angles and the smallest Q available is $\sim 0\cdot 025$ Å^{-1}.

One further important property of the spin-echo technique is its ability to distinguish coherent from incoherent scattering. Most incoherence arises from the neutron spin and since this machine explicitly follows the spin, it is possible to select only the coherently scattered neutrons. In cases where the coherent and incoherent scattering laws differ, this property is a great advantage. At low Q also the coherent scattering law has high intensity for macromolecular systems conveniently increasing the signal-to-noise ratio.

4. MOTION OF SIDE CHAINS

4.1. Inelastic Scattering from Torsional Vibrations

Examination of eqns (10) and (11) shows that large amplitude vibrational modes involving nuclei with large cross-sections will give intense peaks in the inelastic neutron spectra. The incoherent scattering from hydrogen is an order of magnitude larger than that from other nuclei (see Table 1). Torsional vibrations of methyl or phenyl side-chains of polymers with their large amplitude are, therefore, good candidates for neutron investigation, especially since these bands are often weak and difficult to identify from infrared and Raman spectra. Deuteration of the side group where this is possible reduces the cross-section by a factor of around twenty and the peak drops out of the spectra confirming the assignment. Methyl torsion motion has been investigated for poly(propylene oxide),[41] poly(methyl methacrylate),[42] polypropylene,[40, 46–48] poly(dimethyl siloxane)[42, 43, 49] poly(1-alanine),[44] polyacetaldehyde,[45] poly(4-methyl pentene-1),[42] polyisobutene[42] and poly(vinyl methyl ether),[42] and a brief investigation of the phenyl torsion in polystyrene is mentioned in reference 40. In many of these examples, partial deuteration of the molecule has allowed an unambiguous assignment of the torsional mode. In the cases of poly(dimethyl siloxane) and polypropylene a more complicated situation has emerged, in the former case, because deuteration reduces all low frequency modes and is therefore unhelpful and in the latter, because of complications arising from the crystalline structure. These two examples illustrate both the advantages and the problems of neutron scattering spectroscopy.

Henry and Safford[43] used a time-of-flight spectrometer to observe the inelastic scattering from poly(dimethyl siloxane) (PDMS) under various degrees of crosslinking and filler. A broad peak at 160 cm^{-1} for the room temperature sample was assigned to the methyl group motion which was considered to be 'almost free rotation' because of the breadth of the peak. Allen et al.[42] observed a similar peak at 165 cm^{-1}, made the same assignment, and calculated the barrier to rotation assuming a 3-fold potential well. This barrier is about $6 \cdot 8 \text{ kJ mol}^{-1}$ which is about $3 kT$ at room temperature, and argues against the rotation being completely free though it will certainly be fast. A free rotation, being non-quantised gives no inelastic δ functions in eqns (10) and (11) and the elastic peak becomes broadened. Thus $\partial^2 \sigma / \partial \Omega \partial E_f$ (and $S(\mathbf{Q}, \omega)$) are broad distributions centred around $\hbar \omega = 0$. The time of flight cross-section, however, $\partial^2 \sigma / \partial \Omega \partial \tau$, is related to $\partial^2 \sigma / \partial \Omega \partial E_f$ by the factor τ^3 (eqn (25)). This factor distorts the spectrum towards $\hbar \omega \neq 0$ and can even give the appearance of

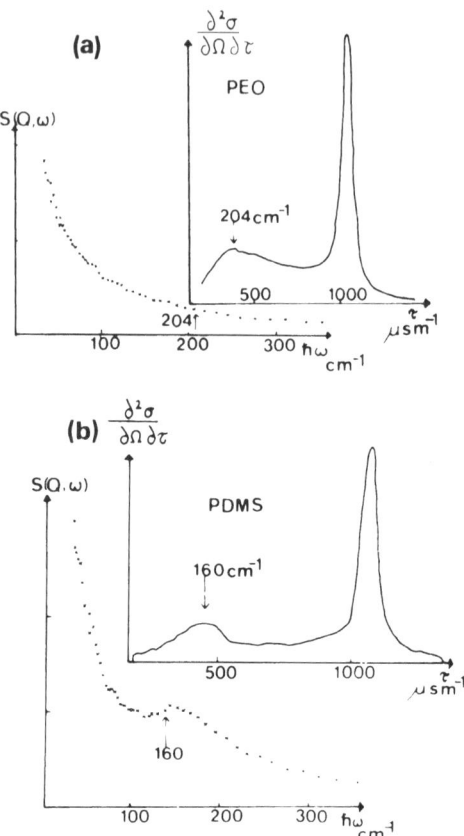

FIG. 6. Time-of-flight spectra $\partial^2\sigma/\partial\Omega\,\partial\tau$ and corresponding scattering laws $S(\mathbf{Q},\omega)$ for: (a) poly(ethylene oxide) (PEO) melts; and (b) poly(dimethyl siloxane) (PDMS).

an inelastic peak. An example of this effect is shown in Fig. 6a where the spectrum of molten poly(ethylene oxide) for which $S(\mathbf{Q},\omega)$ is a smooth function centred at $\hbar\omega = 0$ shows a peak in the time-of-flight spectrum around 200 cm^{-1}. Figure 6b shows the same pair of functions for PDMS. (Both the time-of-flight spectra were obtained at 45° scattering angle with $\lambda_i = 4\cdot2$ Å). In this case, the peak at 160 cm^{-1} is clearly still visible when the τ^3 effect is removed and corresponds to a genuine inelastic feature. Henry and Safford did not attempt to present their data converted from $\partial^2\sigma/\partial\Omega\,\partial\tau$ to $\partial^2\sigma/\partial\Omega\,\partial E_f$ and fitted a calculation for free rotation to the time-of-flight data. The confusion about freedom of rotation of the methyl group in PDMS was extended in a more recent paper where Amaral *et al.*[49] using total scattering cross-section measurements arrive at a barrier to rotation of less than 1·6 kcal mol^{-1}. This measurement, by its nature, is much less accurate than the direct inelastic measurement of the torsional transitions. Although these authors imply that the interpretation of the time-of-flight inelastic spectra in terms of free or hindered rotation is a matter of choice, the above discussion shows that removal of the τ^3 effect leaves no ambiguity. The breadth of the inelastic peak is not due solely to the nature of the methyl rotation itself, but also to the very fast backbone motions in PDMS which broaden both elastic and inelastic δ functions in eqn (11).[50] These motions will be discussed further in Section 5, but it is interesting to note that the inclusion of fast rotations of backbone segments might well account for the low apparent barrier to methyl rotation given by the total scattering experiment. Moreover, many of the effects of filler, of crosslinking and of temperature in narrowing the methyl peak, which were interpreted by Safford and Henry in terms of their effects on the methyl rotation itself, may also be attributed to the effects of slowing down the backbone motion. Finally, very recent measurements of NMR spin lattice relaxation frequencies as a function of temperature for the relaxation mode associated with the methyl motion arrive[51] at an activation energy E_a of about 6·7 kJ mol^{-1}. This is in excellent agreement with the barrier to rotation calculated from the neutron inelastic peak.

The problem with polypropylene has rather different origins. It was one of the first polymers investigated by inelastic neutron scattering.[46] The density of states for the phonon lattice modes was in fairly good agreement with calculation and will be discussed in Section 5, but the assignment of the methyl torsional mode was based on an earlier infrared assignment of a very weak band at 200 cm^{-1}. There is no sign of a peak in the neutron spectrum at this position and this led to the idea that the peak due to the torsional modes would be so broad and its Debye–Walker factor so large that it would be virtually undetectable.[52] Subsequent measurements on poly(propylene oxide),[41] however, did show one clear intense peak which could be unambiguously assigned to the methyl torsion by its total absence in the spectrum of the CD$_3$ analogue. Although broad, the peak was certainly not weak—as it should not be because of the large amplitude hydrogen motion. Calculation[48] of the hydrogen amplitude weighted density of states for polypropylene show that the methyl torsion should be overwhelmingly the most intense peak in the low frequency region. The density of states[40, 46–48] does show a broad, fairly intense, band centred around 230 cm^{-1} which, on deuteration of the remaining hydrogens in the molecule, gains intensity relative to the rest of the spectrum.[48] A reassignment from 210 cm^{-1} to 230 cm^{-1} must be made. However, the peak is nothing like as intense as the torsional mode in poly(propylene oxide), nor as the calculations of $g(\omega)$ suggest it should be. Investigation of an oriented sample subsequently showed strong dispersion effects for this mode, unexpected for a methyl torsion.[48] The experimental results are shown in Fig. 7. A highly stretch-oriented sample was prepared, and when the **Q** vector pointed along the chain axis, the torsional mode was relatively

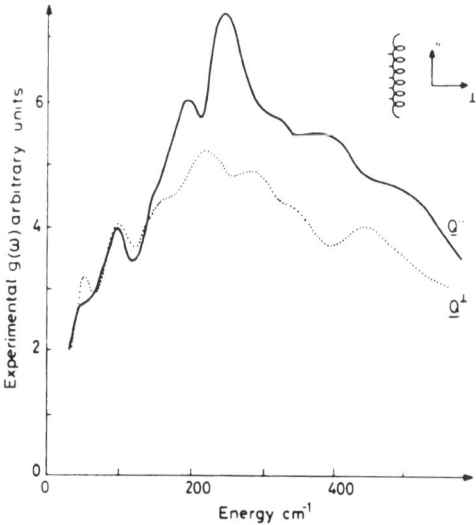

FIG. 7. Experimental density of states, $g(\omega)$ for stretch-oriented polypropylene measured with **Q** parallel and perpendicular to the helical axis. (Reprinted with permission from *Polymer*.[48])

intense and sharp around 240 cm^{-1}. When **Q** was perpendicular to the axis the band was much broader and apparently shifted to 220 cm^{-1}. In this molecule the crystalline form is a helix with the methyl groups sticking out sideways. The explanation of the data in Fig. 7 is that coupling between chains may be enhanced by the methyl groups on neighbouring chains leading to a large frequency dispersion for the methyl torsion centred around 220 cm^{-1} in the direction perpendicular to the helical axis and a small dispersion around 240 cm^{-1} for the parallel direction. In unoriented samples, a broad feature around 230 cm^{-1} results.

These two examples illustrate the importance of the large incoherent cross-section of hydrogen in assigning side chain modes for polymers, together with the necessity of careful interpretation of the time-of-flight spectra which ought to be transformed to $g(\omega)$ if possible. The directional properties of the neutron energy transfer arising from the sizeable momentum transfer are also important. Another example of the use of careful sample–neutron orientation will be discussed in Section 5.

4.2. Rotational Motion of Side Groups

The rotation of a side group between the positions of the potential minima gives rise to a broadened component underneath the elastic scattering (see eqn (16)). The width of this quasielastic component is governed by the rotational frequency. Separation of the elastic from the quasielastic components depends crucially on the incident energy spread (the resolution of the apparatus), and on the shape of the resolution function. To some extent, since the rotational motion is an activated process, its frequency can be adjusted to match the available resolution by changing the sample temperature. However, as a polymeric sample is heated above T_g motion of the main chain will broaden the elastic component (see Section 5), and make separation that much more difficult. For this reason it is preferable to observe side chain rotations in samples at around or below their glass transition temperatures. Although there has been extensive study of quasielastic scattering from rotational motion in plastic and liquid crystals,[53] there has, to date, been only limited work on polymeric systems.

The glass transition of poly(propylene oxide) is 200 K and at this temperature the rotational frequency is around 10^9 Hz demanding the high resolution of the back-scattering spectrometer. This spectrometer has a Lorentzian-shaped resolution function, making separation of the quasielastic component difficult. Nevertheless, comparison with a sample in which the methyl group had been deuterated showed clearly the presence of a quasielastic component[54] and allowed an estimate of the rotational frequency to be made at a number of temperatures. This temperature variation corresponds to an activation energy E_a of 17 kJ mol^{-1} which is in reasonable agreement with the value of $E_a = 15·9$ kJ mol^{-1} estimated from NMR[55, 56] and mechanical relaxation[57] and the value of 13·8 kJ mol^{-1} determined[41] for the depth of the torsional potential well V_3. A direct comparison of E_a with V_3 should, however, take account of quantum mechanical tunnelling effects which lead to observation of an apparently lower barrier to rotation. Stejskal and Gutowsky[58] calculated the average tunnelling frequency and Eisenberg and Reich[59] estimated the effects of tunnelling on the activation energies for methyl rotation in poly(methyl methacrylate). In this case, there was some slight improvement in the agreement between E_a and V_3, but for poly(propylene oxide) inclusion of tunnelling effects only widens the apparent gap between E_a and V_3.[41] This may reflect on several points ranging from the precision with which the torsional peak can be assigned, the form assumed for the potential well in order to calculate V_3 from the torsional frequency and the applicability of the tunnelling calculations themselves.

Poly(methyl methacrylate) (PMMA) is of some interest in this context because not only is it glassy to higher temperatures, but there are two methyl groups in the monomer unit with very different torsional frequencies and some evidence that the crystal structure strongly affects at least one of these torsional barriers.[42, 58, 62, 63] The normal preparation of PMMA produces predominantly syndiotactic sequences and for these samples, the α-methyl torsional band occurs at about 350 cm^{-1} yielding a barrier $V_3 \sim 34$ kJ mol^{-1}. In earlier work, this band was assigned by comparing the α chloro analogue with PMMA itself,[42] but deuteration of this group has recently reconfirmed this assignment.[60, 61] In a predominantly isotactic sample this barrier is reduced to 23 kJ mol^{-1}. The ester methyl group has a torsional band at rather lower energy (around 100 cm^{-1}) leading to a value for V_3 of between 3 and 10 kJ mol^{-1}. The band is difficult to assign with more precision because it occurs in a region of other backbone motions which are also reduced in intensity in the neutron spectra after deuteration of the methyl group.[42, 60, 61] From the inelastic spectra, it is not possible to determine the effect of tacticity on this barrier. These barriers are different enough so that even without the aid of specific deuteration, it is possible to separate the effects of their rotational motion in the quasielastic spectra for samples below T_g. At room temperature, the α-methyl group is rotating at less than 10^9 Hz while the ester methyl moves much faster at 10^{11} Hz. Specific deuteration reduces the elastic scattering from the other hydrogen nuclei (immobile in the glass on this time scale apart from vibrations which are such high frequency they average out). Using the time-of-flight spectrometer, IN5, described in Section 3, with its resolution adjusted to match the appropriate rotational frequency, and the back-scattering spectrometer for the slowest rotation, it was possible to separate the elastic and quasielastic contributions for both methyl groups and to extract the rotational frequency ν_{rot} and the elastic incoherent structure factor EISF (see eqn (16)). Figure 8 shows $S(\mathbf{Q}, \omega)$ in the quasielastic scattering region from a sample of predominantly syndiotactic PMMA in which all the hydrogens apart from the ester methyl have been deuterated. The time-of-flight spectrometer has an approximately triangular resolution function which makes the two-component nature of the

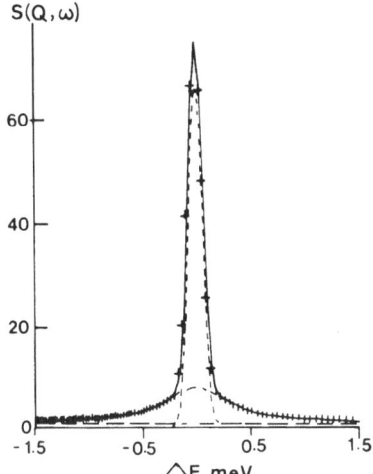

FIG. 8. Quasielastic scattering from poly(methyl methacrylate) at 25°C showing the rotational motion of the ester methyl group as a broadened component under a resolution broadened δ-function. $Q = 2·06 \text{ Å}^{-1}$.

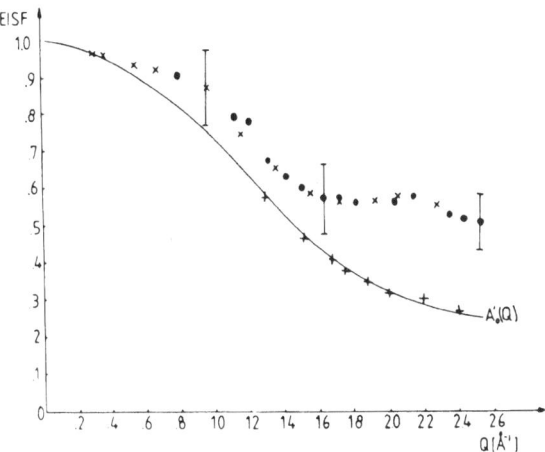

FIG. 9. EISF for the ester methyl rotation in poly(methyl methacrylate). The theoretical calculation including corrections for elastic incoherent scattering $A_0'(Q)$ is shown as a solid line (+); (×), IN5 data (δ fn + single Lorentzian); ●, IN6 data (3-fold jump).

158   J. S. HIGGINS

scattering evident. The fits to the data of a δ-function and a broadened Lorentzian, both resolution broadened, are also shown. Figure 9 shows the value of $A_0(Q)$ as a function of Q extracted using two different methods of separating the data and from two different experiments. The curve $A'_0(Q)$ is the calculated EISF for a rotating methyl group adjusted for the extra incoherent scattering from the remaining nuclei in the molecule. There is a mismatch between the calculated and experimental values. However, the coherent scattering is not inconsiderable from this sample in which five out of the eight hydrogen nuclei have been deuterated. The X-ray scattering from this polymer shows[64] peaks at around $Q = 0.8$ Å^{-1} and 2.2 Å^{-1}. The neutron diffraction from the same sample as that used for the rotational measurements also showed peaks around these values at the points of maximum deviation of the data from the calculated $A'_0(Q)$. Correction of

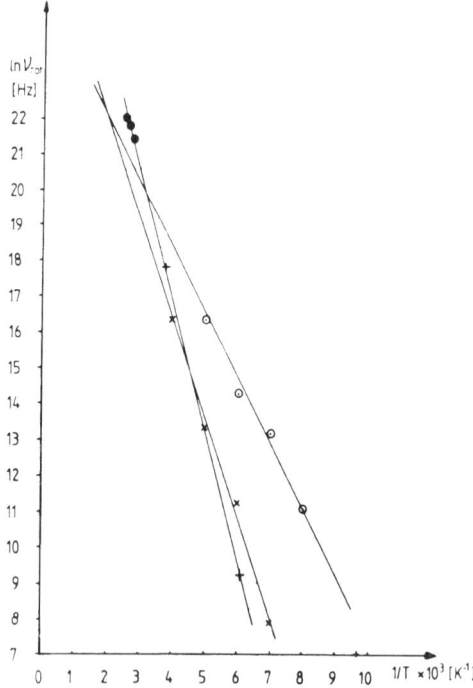

FIG. 10.   Arrhenius plots of neutron and NMR frequencies for the α-methyl rotation in polymethyl methacrylate. ×, syndiotactic PMMA (NMR data); ⊙, isotactic PMMA (NMR data); +, atactic PMMA (NMR data); ●, atactic PMMA (IN10 data).

the data for this coherent elastic contribution brings the data into excellent agreement with the calculations as shown by the crosses in Fig. 9. Since $A_0(Q)$ is the Fourier transform of the rotational volume, a modification of the proposed rotational motion to achieve the shape of the experimental $A_0(Q)$ would demand a reduction in the rotational radius—which is not possible for a methyl group rotating around its C axis unless we imagine shortening the C—H bonds or drastically changing the bond angles!

TABLE 3
BARRIERS TO ROTATION IN kJ mol^{-1} FOR PMMA

Sample	$E_a \rightarrow V_3$ neutrons	tunnelling	$E_a \rightarrow V_3$ NMR	tunnelling	V_3 neutron inelastic measurements
α methyl predominantly syndiotactic	29·1 $\rightarrow$ 31·1				33
α methyl pure syndiotactic			23·3 $\rightarrow$ 30		33 33
α methyl pure isotactic			15·8 $\rightarrow$ 22·6		23
Ester methyl predominantly syndiotactic	7·6 $\rightarrow$ 8·8		10·1		3·2 to 10·5

Figure 10 shows a plot of the rotational frequencies obtained from neutron quasielastic measurements on the α-methyl rotation in a predominantly syndiotactic sample, together with NMR relaxation values.[58, 62, 63] Excellent agreement is obtained in the values of E_a from NMR and neutron experiments. The rotational barrier for the ester methyl rotation is so low that NMR measurements only just observe it and there is difficulty in obtaining a precise value for E_a. The neutron quasielastic scattering results are again unambiguous and give a value for E_a in good agreement with estimates from NMR relaxation and from the neutron measurements of the frequency of the torsional vibration. Table 3 compares values of E_a from NMR and neutron experiments with values of V_3 from inelastic measurements; (when converting E_a to V_3 adjustment has been made for the tunnelling effects discussed above). Completion of the neutron experiments will eventually allow values of E_a to be assigned for both groups in both stereo isomers with much better precision than that allowed by calculation from the rather poorly resolved torsional frequencies (and better than NMR estimates for the very low ester methyl barrier). Preliminary data[60, 61] indicate that the ester barrier also may be affected by tacticity.

5. MOTION OF THE POLYMER BACKBONE

5.1. Crystalline Samples–Phonon Modes

The observation of travelling waves in polymer crystals as described in Section 2.2.1 offers the attractive possibility of obtaining both the inter- and infra-chain force constants. The drawback is the near impossibility of obtaining single crystals of polymers. Most polymers are only semi-crystalline with small crystallites embedded in an amorphous matrix. The scattering from such an unoriented sample essentially averages out much of the directional information in eqn (11). However, in the coherent scattering, for each value of ω there is a single allowed Q value and the ω–Q pairs must obey the dispersion relationship. Thus the dispersion curve of a particular vibrational mode may be picked out from a series of time-of-flight spectra at different scattering angles or followed on a triple-axis spectrometer by tracing the way a peak moves in ω–Q space. The coherent scattering from polytetrafluorethylene[65] and deuterated polyethylene[66] is strong enough to allow the acoustic modes to be analysed in this way and compared with calculated dispersion curves. For hydrogenous polymers, however, the incoherent scattering dominates and only the H amplitude weighted density of states, $g(\omega)$ is observed.[46, 67] This may be compared with a calculated curve (see eqns (12) and (13)) but such comparison does not provide a very stringent or direct test of the force constants used.[52, 67] If the polymer samples are stretch-oriented, then some of the directional information is recovered.

Figure 11 shows the density of states $g(\omega)$ for stretch-oriented polyethylene determined using a triple-axis spectrometer so that in one case, the Q vector was oriented always along the chain axis (longitudinal) and in the second, Q was perpendicular to this axis (transverse).[68] The two curves have been normalised at 190 cm^{-1}. The two main peaks at 525 and 190 cm^{-1} have been assigned respectively to the longitudinal stretch bend (accordion) mode of the C—C—C skeleton and to the out-of-plane torsion of the methylene groups about the C—C bond. The disappearance of the 525 cm^{-1} band from the $Q^{\parallel}$ spectrum confirmed its assignment to the

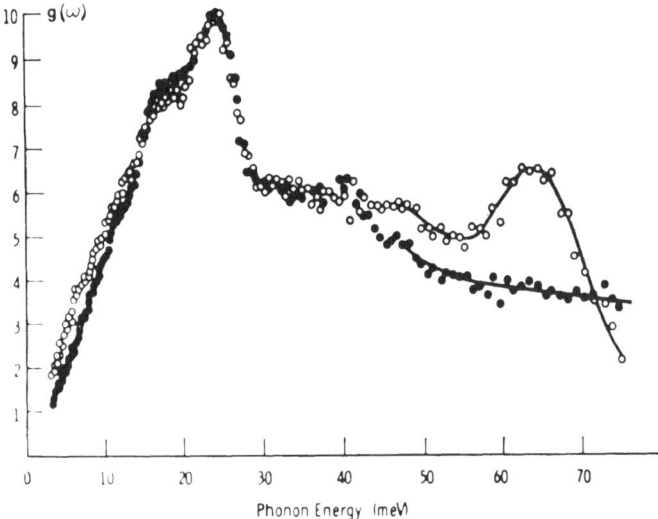

FIG. 11. Amplitude weighted direction vibrational frequency spectra for a stretch-oriented sample of polyethylene at 100 K. (Reprinted with permission from *J. Chem. Phys.*[68]) ○, longitudinal $g(\omega)$; ●, transverse $g(\omega)$.

longitudinal mode—in this configuration there is no way for the neutron to excite the motion. The fact that the transverse mode shows up in both spectra (though without the normalisation mentioned, it is, in fact, reduced in intensity by a factor two in the $Q^{\parallel}$ spectrum) is caused by crystalline field mixing of the modes.

Stretch-orientation of coherently scattering samples also helps the measurement of dispersion curves and these have been compared with calculations for deuterated polyethylene[69] and PTFE,[70] both stretch-oriented. In only one case,[71] a large single crystal was grown (of poly-oxymethylene) and a number of dispersion curves reported. Where it has been possible to measure the dispersion of the interchain modes, the force constants obtained provide unique information specific to the neutron technique. However, as can be seen, the amount of experimental data is limited by the difficulties of working with polycrystalline, at best partially-oriented samples.

5.2. Polymer Solutions

As explained in Section 2.2.3, the type of main chain motion observed in quasielastic scattering from a polymer solution depends on the frequency resolution and Q-range of the experiment. Early measurements on polymer solutions were restricted by the resolution of the available time-of-flight apparatus to frequencies above 10^{11} Hz. These frequencies are only observed at relatively large Q values and correspond to motion over distances of the order of 1–2 Å—local bending and rotation of single bonds. The high resolution back-scattering and spin-echo spectrometers (see Sections 3.4 and 3.5) have pushed the limiting frequency range down several orders of magnitude to around 10^7 Hz. Motion over distances up to 30 Å is now accessible. In this Q-range, the coherent signal from a large

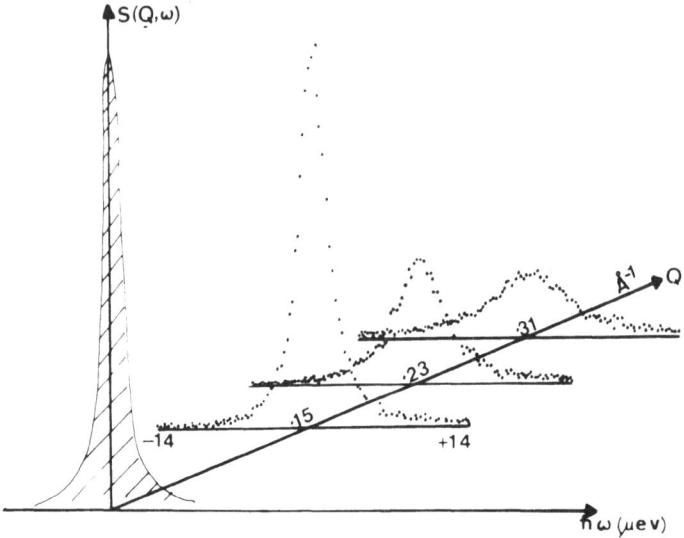

FIG. 12. $S(\mathbf{Q}, \omega)$ at a number of Q values measured using the IN10 back-scattering spectrometer for a 3% solution of perdeutero polystyrene in CS_2 at room temperature. The curve at zero Q is the resolution function of the spectrometer.

object such as a polymer molecule becomes very intense provided there is a difference in scattering length density between the molecule and its surroundings.[50, 72] Figures 12 and 13 show $S(\mathbf{Q}, \omega)$ and $S(\mathbf{Q}, t)$ respectively for a 3% solution of perdeutero polystyrene in CS_2; Table 1 shows there is a large difference in scattering length densities between these molecules. Figure 12 contains data obtained at ILL on the back-scattering spectrometer at three different values of Q. As Q increases, the peak broadens, and decreases in intensity to a limiting value as the coherent scattering falls off. The scattering from CS_2 alone has been subtracted, so at high Q the residual intensity is dominated by the incoherent scattering from the

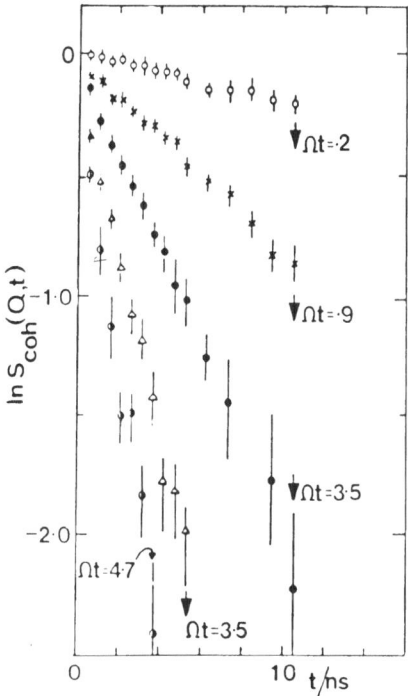

FIG. 13.   Correlation functions at a number of Q values for the same sample as in Fig. 12 measured using the spin-echo spectrometer IN11. Q values: ○, 0·026; ×, 0·05; ●, 0·08; △, 0·11; ◑, 0·13, Å^{-1}. (Reprinted with permission from *Macromolecules*,[76] Copyright (1981) American Chemical Society.)

polymer. The curve depicted at zero Q represents the resolution function of the apparatus (not normalised to the other curves). As can be seen the quasielastic scattering at the lowest Q value ($Q = 0·15$ Å^{-1}) is very close in shape to the resolution function (measured by the scattering from vanadium) and it is difficult to extract the width of $S(\mathbf{Q}, \omega)$ from the resolution broadened spectrum. In the example shown in Fig. 12, the resolution has a value of $\Delta\omega = 0·6\ \mu$eV and the value for $S(\mathbf{Q}, \omega)$ itself is only 0·7 μeV. The slowest variation of $\Delta\omega$ with Q quoted in Table 2 is Q^2 so that decreasing Q to 0·03 Å^{-1} would reduce $\Delta\omega$ by a factor of twenty five!

For polymer solutions the $\mathbf{Q}$ variation is normally faster than Q^2 so it is easy to see why experimentalists interested in observing long-range correlated motion of polymers have been continually chasing the highest resolution apparatus available. At present, this is the spin-echo spectrometer at ILL.

In Fig. 13, the lowest Q data are measured at $Q = 0.026$ Å^{-1}, and it is still possible to detect the inverse correlation time (about 2×10^7 s)—the slope of this logarithmic plot. As described in Section 3.5, this apparatus makes it possible (by explicitly following the neutron spin) to separate the coherent from the incoherent scattering. The curves in Fig. 13 correspond to purely coherent scattering.

Equations (19) and (20) show that simple Brownian motion would give Lorentzian curves in Fig. 12 (since the resolution function is approximately Lorentzian also, this form is preserved even after resolution broadening) and simple exponentials, i.e. straight lines, in the logarithmic plots of Fig. 13. Internal motion in the form of Rouse or Zimm modes would give deviations from these shapes. Close examination of the data such as in Fig. 13 does show small deviations from the Lorentzian shape,[73, 74] but because model curves have to be convoluted with the resolution function before comparison with the data, it is difficult to make more definite conclusions than the generalisation that deviations exist. In the same way, the values of $\Delta\omega$ extracted are dependent on the model† chosen[73, 74] and at the higher Q values will depend on some mixture of coherent and incoherent scattering. Table 2 shows that the numerical coefficients of the various results change for $\Delta\omega$ obtained from coherent and incoherent scattering, but the Q variation is the same in each case. Using measurements of $S(Q, \omega)$, despite the drawbacks mentioned, it was possible to show that the Q dependence of $\Delta\omega$ was faster than Q^2 and that Zimm modes were observed in the neutron experiments.[73-75]

The ability to observe $S(Q, t)$ directly (and to separate the purely coherent scattering) using the spin-echo technique has obvious advantages. The Fourier transform of a convolution of 2 functions is a product of these functions so that in time domain experiments, the resolution is simply divided out of the experimental spectrum and this can be done involving no assumptions about the model needed to describe the scattering. As discussed in Section 2.2.3, Rouse or Zimm modes would produce deviations from the simple exponentials in eqn (20), but very little deviation from straight lines is visible in the corrected data in Fig. 13. This is explained when the values of the normalised time Ωt shown on the figure, are examined. For the low Q curves Ωt is still very small at the limit of the experimental time scale. However, in Fig. 14 where the theoretical correlation functions calculated by Akcasu et al.[20] are plotted in the same

† It was found important when deciding on the best fit model to describe the data to consider results obtained with widely different resolution function (i.e. different spectrometers).[73, 74, 85]

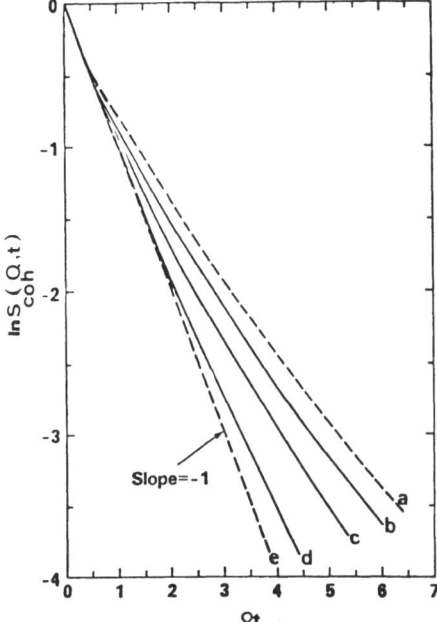

FIG. 14. Theoretical correlation functions for various values of Qa. (Reprinted with permission from *Macromolecules*,[76] Copyright (1981) American Chemical Society.) a, $Qa \leq 1$; b, $Qa = 2$; c, $Qa = 4$; d, $Qa = 6$; e, $Qa > 10$.

way as the experimental data in Fig. 13, but against normalised time, it can be seen that very little deviation from straight line behaviour would be observed for $\Omega t < 1$. The curves in Fig. 14 are plotted for different values of Qa where a is the Rouse spring length (as yet an unknown parameter). For a given a, as Q increases, the curves return to straight line behaviour characteristic of local Brownian motion—this corresponds to the upper horizontal region in Fig. 1. It is these higher Q correlation functions that stretch further out in Ωt in Fig. 13, but these are now expected to be linear as is observed.

Data such as that shown in Fig. 13 for polystyrene solutions have been analysed for a number of polymers in different solvents.[76] Values of Ω were extracted from the observed $S(Q, t)$ functions by fitting the de Gennes[18, 19] calculated scattering laws where appropriate (and remembering $\Omega = \sqrt{2}\theta/t$) and using the initial slope of the logarithmic plots otherwise. Only the very flexible polymer, poly(dimethyl siloxane), showed approximately Q^3 behaviour as described in eqn (21). Both polystyrene and polytetrahydrofuran showed intermediate behaviour characteristic of the region around $Qa = 1$ in Fig. 1. Examination of the axes in Fig. 1 shows that all parameters are fixed by the experimental conditions except a. If plots of Ω/Q^2 vs Q for each solution (with a given solvent and temperature so T and η are fixed) are slid diagonally across the master curve in Fig. 1 until the best fit is obtained, the shift factors give the values of a. A combined curve for all the data in one solvent can be built up by normalising each set of data by its a value. Figure 15 shows such a set of data for polystyrene, polytetrahydrofuran and poly(dimethyl siloxane) in

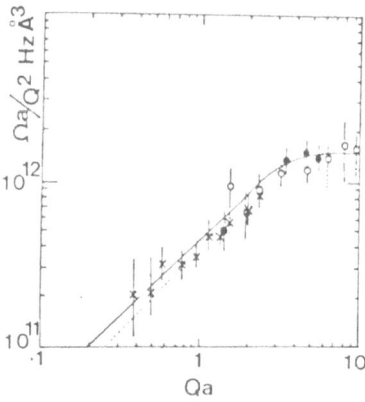

FIG. 15. $\Omega a/Q^2$ against Qa for three polymers in 3% solutions in C_6D_6 at room temperature. ×, poly(dimethyl siloxane); ○, polytetrahydrofuran; ●, polystyrene. The solid curve is the theoretical calculation for a good solvent (see Fig. 1 and ref. 20). $\eta_s = 0.56$ cp; $T = 30°C$. (Reprinted with permission from *Macromolecules,* [76] Copyright (1981) American Chemical Society.)

3% solutions in C_6D_6 at 30°C. The a values used were 55, 45 and 16 Å, respectively. The interpretation of a as the length of the imaginary Rouse spring unit should not be taken too literally. It is perhaps better to think of a as giving a measure of the distance scale below which this simplified model must be replaced by calculations which take account of the local structure of the polymer molecule. Preliminary calculations of this sort[77] for a polyethylene chain do show the return towards the Q^2 dependence of Ω observed in practice. Techniques such as high frequency viscoelastic measurements[7, 78] test the frequency limits of the Rouse and Zimm models. The frequency at which the horizontal portion of the curve in Fig. 1 begins for each of the polymers mentioned, marking the limit of Rouse–Zimm modes, ranges from 10^8 s for polystyrene to 4×10^9 s for poly(dimethyl siloxane). After correcting for the high viscosity of the solvents usually used in viscoelastic measurements in order to reach these high frequency limits, we find that the limit for polystyrene is around 10^8 Hz for these measurements as well. The neutron experiment, however, has explicitly brought the spatial limitations of the model into consideration. When detailed comparison is made with techniques such as ^{13}C NMR, dielectric relaxation and fluorescence depolarisation which also give high frequency information, the neutron results, with their explicit spatial information help the interpretation of the observed frequency in terms of motion of different length sections of the molecule.[76]

It is not expected[76] that new neutron techniques will improve the frequency or Q ranges available in the foreseeable future so these experiments map out the accessible range as the upper end of the Rouse–Zimm

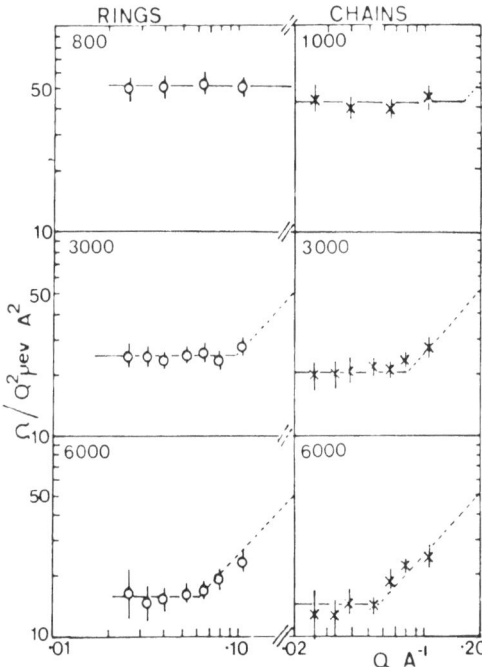

FIG. 16. Ω/Q^2 vs Q for a sequence of low molecular weight poly(dimethyl siloxane) molecules of ring and chain form in 3% solution in C_6D_6 at 30°C.

R_g values   800: <5   ⎫   1 000: <6   ⎫
    Å   3 000: 9·54 ⎬ rings   3 000: 13·1 ⎬ chains
        6 000: 14·3   ⎭   6 000: 19·9   ⎭

frequency range. Within this range, further exploration is taking place on the effect of solvent, concentration and temperature.[79-81]

Figure 16 shows what happens if molecules are made small enough so that the lower horizontal range (limited by $QR_g = 1$) falls within the Q range of the neutron experiment. Ω/Q^2 is plotted against Q for a number of low molecular weight linear and cyclic poly(dimethyl siloxane)s in 3% solution in C_6D_6. The diagonal dotted line corresponds to eqn (21) with the appropriate coefficient from Table 2. The horizontal portion corresponds to centre of mass diffusion where $\Omega = \Theta/t = DQ^2$ so that the diffusion coefficient D can be extracted from the data. Results are in very good agreement with data on chain dimensions from elastic small-angle neutron scattering measurements.[82] Such small molecules cannot be observed in photon correlation spectroscopy conventionally used to measure diffusion of polymer molecules so this spin-echo technique conveniently extends the range of such measurements where necessary. It is interesting to note that for small stiff molecules, the two limits $QR_g = 1$ and $Qa = 1$ will approach each other and the Q^3 region may disappear completely!

5.3. Polymer Melts

Figure 17 shows the values of Ω for samples of deuterated polytetra-hydrofuran containing a small percentage of the hydrogenous polymer.[25] The difference in scattering length for H and D gives a large coherent scattering pattern from the differentiated chains[72] so that the dynamics of single molecules in a melt environment can be observed. Also shown in Fig. 17 are the results for the same polymer in dilute solution in CS_2. The data in Fig. 17 have been extended to higher Q by including results for $\Delta\omega$ obtained from measurements on the back-scattering spectrometer. Allowance has been made for the inevitable contamination of $S(Q, \omega)$ by incoherent scattering at higher Q. In a melt, the hydrodynamic interactions in solution, characteristic of the Zimm treatment, are screened out by the surrounding chains and provided the distances explored do not include entanglements, the pure Rouse behaviour would be expected to give a Q^4 dependence of Ω or Θ/t as shown in eqn (22). Ω certainly does show a faster Q dependence of the bulk data than the solution data in Fig. 17, though neither of the slopes are at the limiting values of Q^3 and Q^4.

The smallest Q value corresponds to a distance of 30 Å. Recent discussions describe the motion of polymer molecules in a melt in terms of reptation[23, 83] along a 'tube' formed by the effect of the surrounding chains. Graessley[24] has recently calculated the diameter of this tube from experimentally measured quantities for several polymers and found that it is

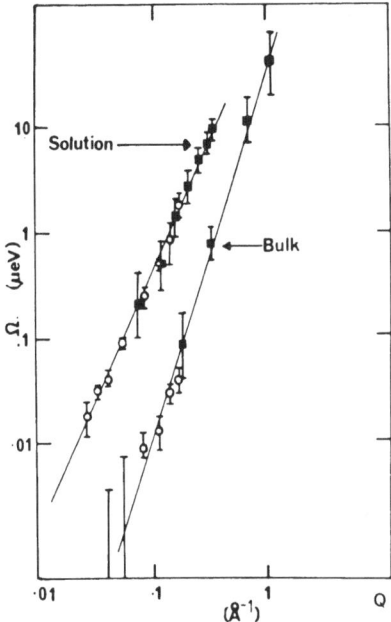

FIG. 17. Ω against Q for 3% perdeutero polytetrahydrofuran in CS_2 and for 3% polytetrahydrofuran in perdeutero polytetrahydrofuran melt. $\circ$, data obtained from $S(Q, t)$ from the spin-echo spectrometer; $\blacksquare$, data obtained from $S(Q, \omega)$ from the back-scattering spectrometer. (Reprinted with permission from *Polymer*.[25]).

surprisingly large—of the order of 35 Å for polyethylene. It would not be expected, therefore, that entanglement effects would show up over the distance scales explored in the neutron experiments.

Again, the melt data are in the high frequency limit of the Rouse modes, and further analysis must include more detailed consideration of the local structure. It is, however, possible to compare the results from different polymer melts and draw some conclusions about their relative rates of segmental motion.

Results of experiments measuring $S(Q, \omega)$ for the incoherent scattering from a number of polymer melts using the back-scattering spectrometer are summarised in Fig. 18.[84] On this double logarithmic plot, the slope is the power of Q variation. For poly(dimethyl siloxane)[85] this is approximately Q^4, but for all the other polymers it is somewhat less, and for polyisobutylene at 73°C, it is not much faster than Q^2. Polypropylene oxide, polytetrahydrofuran and polyisobutylene all have glass transition

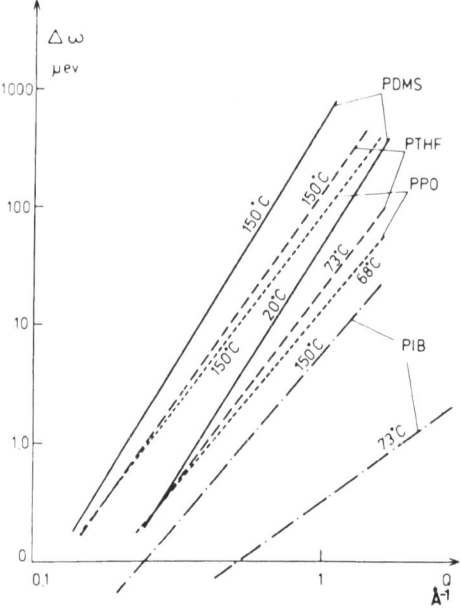

FIG. 18. $\Delta\omega$ against Q plotted logarithmically for poly(dimethyl siloxane) (PDMS), polytetrahydrofuran (PTHF), poly(propylene oxide) (PPO) and poly-isobutylene (PIB) at two temperatures.

temperatures around 200 K. The very much slower motion and stiffer unit (if we interpret a low power Q variation in terms of Fig. 1 leading to a long spring unit a) for polyisobutylene is somewhat surprising if the glass transition is interpreted in terms of the freezing out of local segmental motion. Poly(dimethyl siloxane) does have a much lower T_g than the other polymers, so its faster motion and more flexible structure (in terms of a short a value) fits this interpretation. The Q range in Fig. 18 extends much higher than that in Figs 15 and 16 and the persistance of power laws greater than Q^2 to such values is somewhat puzzling in terms of Fig. 1 since it is clearly unrealistic to imagine values of a less than 2 or 3 Å. At present it seems the explanation must lie in the fact that the melt can sustain rather low frequency transmitted modes like those described in Section 2.2.1 for crystalline samples. (After all, sound waves are transmitted through a rubber or liquid even though they may be heavily damped). The low frequency density of states spectrum in this case would be indistinguishable from the wings of the $S(Q, \omega)$ arising from conformational modes and lead to an overestimation of $\Delta\omega$. It certainly appears that more careful modelling of all the local motion sustained by the melt polymers will be necessary in order to explain the behaviour of $S(Q, \omega)$ and $\Delta\omega$ at higher Q.

5.4. Networks

When a polymer melt is crosslinked to form a rubber, the macroscopic motion of the molecules is removed and a three-dimensional network is produced. It is by no means clear that on the local scale observed in the neutron experiments the same modifications take place. In fact, experiments on model networks[86] showed no effects of cross-linking until the links were relatively close ($\sim$30 monomer units) together. The question then arises as to whether the resultant slower motion observed as narrower $S(Q, \omega)$ curves is a slower motion of the whole molecule, or a superposition of slower motion of the junctions onto an effectively unaltered Rouse spectrum. Comparison with a sample in which the sections of chains away from junctions had been deuterated showed that it is, in fact, just the section around the junctions which are slowed down.[87] Figure 19 shows data plotted as $\Delta\omega/Q^2$ against Q for such model networks at 70°C. The curve marked ($\bullet$) is for a fully hydrogenous model tri-functional network of poly(propylene oxide) with 30 monomer units between each junction. The points marked ($\times$) are data for an identical network which has been deuterated except for a few units around each crosslink. Finally, the upper set of data marked ($\blacktriangle$) is the value of $\Delta\omega$ from an appropriately weighted difference between the two previous sets of $S(Q, \omega)$ data, and therefore represents the motion of the free chain sections away from junctions. This is indistinguishable from the motion of uncrosslinked poly(propylene oxide) melt at the same temperature indicated by the ($\blacksquare$) symbols in Fig. 19. Calculations[88] of the effect of crosslinking on the Rouse motion of a polymer chain indicate that the Rouse spectrum is unaltered away from the crosslinks and that the crosslinks should reduce their frequency by a factor $2/N$ where N is the functionality of the network. The reduction factor for the trifunctional networks in Fig. 19 is approximately $\frac{1}{2}$ rather than $\frac{2}{3}$ predicted, but this discrepancy is probably not surprising since the neutron experiments only explore the limits of the Rouse model. The calculation assumes that the crosslink points are infinitely far apart, which corresponds to an assumption that a given Rouse mode cannot propagate between the junctions during the time of the experiment. This assumption would seem to be justified for the high frequencies explored by the neutron experiments, but when the junctions become close enough together a modification of even the high frequency Rouse spectrum must be expected.

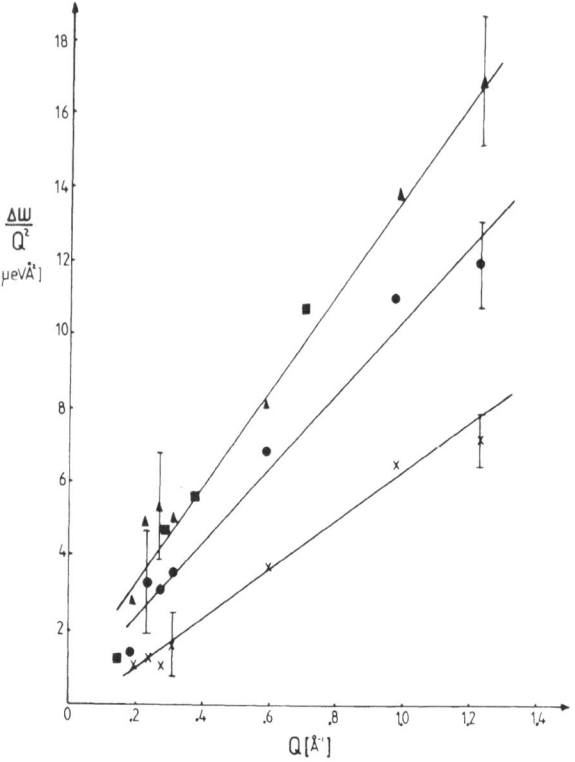

FIG. 19. $\Delta\omega/Q^2$ against Q for model networks of polypropylene oxide. ●, hydrogenous network; ×, partially deuterated network; ▲, difference values as described in the text; ■, values for polypropylene oxide melt at the same temperature (70°C).

6. SUMMARY AND FUTURE WORK

The examples described in detail in Sections 4 and 5 have emphasised both the strengths and the weaknesses of the neutron techniques when applied to observation of polymer motion. Flux limitations will always ultimately frustrate the search for higher resolution so that more common techniques such as IR and Raman spectroscopy or NMR relaxation should always be used in cases where the specific neutron properties offer no particular advantage. One such advantage is the use of deuterium labelling, which not only shifts frequency as in the more conventional methods, but grossly changes the relative intensities of peaks and allows motion of specific groups of molecules or segments of molecules to be conveniently highlighted. The large wave vector associated with the neutron because of its mass leads to specific spatial information unobtainable by other methods. As methods of preparation of stretch-oriented, highly crystalline materials are improved, the information on inter- and intra-chain force constants obtained by tracing phonon dispersion curves will be extended. In a number of cases, the mechanical and NMR relaxations associated with side group motions are so fast around room temperature that their assignment to data has been uncertain. The use of quasielastic scattering to observe this motion not only gives more precise values for the barrier to rotation, but pins down the exact nature of the rotational motion via the elastic incoherent structure factor. The backbone segmental motion in solutions and melts has been shown to be well described in terms of the simple Rouse bead–spring model. Although it is not possible, unfortunately, to explore distances where the entangled nature of the polymer melt comes into play, it is possible to explore the effects of polymer structure and environment on these higher frequency modes. The effect of cross-linking has already been discussed, now experiments are beginning on blends and copolymers, on biological systems which are inevitably very complicated, and on polymers adsorbed on or included within substrates. The effects of distorting the chains by stretching or compressing them will also be of interest here, and it remains to be seen whether coupling of pulsed samples to the new pulsed sources will lead to totally new types of experiments in the next few years.

REFERENCES

1. WILLIS, B. T. M. (Ed.), *Chemical Applications of Thermal Neutron Scattering*, 1973, Oxford University Press, Oxford.
2. KOSTORZ, G. (Ed.), *Treatise on Materials Science and Technology—Vol. 15: Neutron Scattering*, 1979, Academic Press, London.
3. MACONNACHIE, A. and RICHARDS, R. W., *Polymer*, 1978, **19**, 739.
4. ALLEN, G. and HIGGINS, J. S., *Rep. Prog. Phys.*, 1973, **36**, 1073.
5. MCBRIERTY, V., *Polymer*, 1974, **15**, 503.
6. IVIN, K. J. (Ed.), *Structural Studies of Macromolecules by Spectroscopic Methods*, 1976, Wiley Interscience, London.
7. OSAKI, K. and SCHRAG, J. L., *Polym. J.*, 1971, **2**, 541.
8. DUNBAR, J. H., NORTH, A. M., PETHRICK, R. A. and TEIK, P. B., *Polymer*, 1980, **21**, 764.
9. VAN HOVE, L., *Phys. Rev.*, 1954, **95**, 249.
10. HOWARD, J. and WADDINGTON, T. C., *Mol. Spectroscopy with Neutrons*, in:

Advances in Infra-red and Raman Spectroscopy, R. J. H. Clark and R. E. Hester, Eds, Vol. 7, 1980, Heyden, London.

11. BARNES, J. D., *J. Chem. Phys.*, 1973, **58**, 5193.
12. SKOLD, K., *J. Chem. Phys.*, 1968, **49**, 2443.
13. HERVET, H., DIANOUX, A. J., LECHNER, R. E. and VOLINO, F., *J. Phys. (Paris)*, 1976, **37**, 587.
14. BERNE, B. J. and PECORA, R., *Dynamic Light Scattering With Applications to Chemistry, Biology and Physics*, 1976, Wiley & Sons, London.
15. ROUSE, P. JR., *J. Chem. Phys.*, 1953, **21**, 1272.
16. ZIMM, B., *J. Chem. Phys.*, 1956, **24**, 269.
17. PECORA, R., *J. Chem. Phys.*, 1968, **49**, 1032.
18. DE GENNES, P. G., *Physics*, 1967, **3**, 37.
19. DU BOIS VIOLETTE, E. and DE GENNES, P. G., *Physics*, 1967, **3**, 181.
20. AKCASU, A. Z., BENMOUNA, M. and HAN, C. C., *Polymer*, 1980, **21**, 866.
21. AKCASU, A. Z. and BENMOUNA, M., *Macromolecules*, 1978, **11**, 1193.
22. AKCASU, A. Z., BENMOUNA, M. and ALKHAFAJI, S., *Macromolecules*, 1981, **14**, 147.
23. DOI, M. and EDWARDS, S. F., *J. Chem. Soc., Faraday II*, 1978, **74**, 1789, 1802, 1818.
24. GRAESSLEY, W. W., *J. Polym. Sci. Polym. Phys.*, 1980, **18**, 27.
25. HIGGINS, J. S., NICHOLSON, L. K. and HAYTER, J. B., *Polymer*, 1981, **22**, 163.
26. LYNN, J. E., *Contemporary Phys.*, 1980, **21**, 483–500.
27. FENDER, B. E., HOBBIS, I. C. W. and MANNING, G., *Phil. Trans. R. Soc. London*, 1980, **B290**, 657–672.
28. WINDSOR, C., *Pulsed Neutron Scattering*, 1981, Taylor and Francis, London.
29. The Institut Laue–Langevin is owned and run jointly by France, Germany and the UK for the benefit of their scientific communities. For further information, contact the Scientific Secretariat, ILL, BP156X Centre de Tri, 38042 Grenoble, France.
30. STIRLING, W. G., *Neutron Beam Facilities at the HFR*, ILL (see ref. 29).
31. IYENGAR, P. K. in: *Thermal Neutron Scattering*, P. A. Egelstaff, Ed., 1965, Academic Press, London.
32. DOLLING, G. in: *Dynamical Properties of Solids—I*, G. K. Horton and A. A. Maradudin, Eds., 1974, Elsevier North Holland Publ. Co., Amsterdam.
33. BROCKHOUSE, B. N. in: *Inelastic Scattering of Neutrons from Solids and Liquids*, Proc. Symposium IAEA, 1961, p. 113.
34. BRUGGER, R. in: *Thermal Neutron Scattering*, P. A. Egelstaff, Ed., 1965, Academic Press, London.
35. DOUCHIN, F., LECHNER, R. E. and BLANC, Y., ILL Internal Scientific Report, 1973, ITR 26/73 and ITR 12/73.
36. BIRR, M., HEIDEMANN, A. and ALEFELD, B., *Nucl. Inst. Methods*, 1971, **95**, 435.
37. MEZEI, F., *Z. Phys.*, 1972, **255**, 146.
38. HAYTER, J. B. in: *Neutron Diffraction*, H. Dachs, Ed., 1978, Springer Verlag, Berlin.
39. (a) DAGLEISH, P., HAYTER, J. B. and MEZEI, F. in: *Neutron Spin-Echo*, F. Mezei, Ed., 1980, Physics 128, Springer Verlag, Berlin. (b) HAYTER, J. B., *ibid*.
40. WRIGHT, C. J., Chapter 3 in ref. 6.
41. ALLEN, G., BRIER, P. N. and HIGGINS, J. S., *Polymer*, 1972, **13**, 157.

42. ALLEN, G., WRIGHT, C. J. and HIGGINS, J. S., *Polymer*, 1975, **15**, 319.
43. HENRY, A. W. and SAFFORD, G. J., *J. Polym. Sci.*, *A–2*, 1969, **7**, 433.
44. DREXEL, W. and PETICOLAS, W. L., *Biopolymers*, 1975, **14**, 715.
45. LONGSTER, G. F. and WHITE, J. W., *Mol. Phys.*, 1969, **17**, 1.
46. SAFFORD, G. J., DANNER, H. R., BOUTIN, H. and BERGER, M., *J. Chem. Phys.*, 1964, **40**, 1426.
47. YASUKAWA, T., KIMURA, M., WATANABE, N. and YAMADA, Y., *J. Chem. Phys.*, 1971, **55**, 983.
48. TAKEUCHI, H., HIGGINS, J. S., HILL, A., MACONNACHIE, A., ALLEN, G. and STIRLING, G. C., *Polymer*, 1982, **23**, 499.
49. AMARAL, L. W., VINHAS, L. A. and HERDADE, S. B., *J. Polym. Sci., Polym. Phys.*, 1976, **14**, 1077.
50. HIGGINS, J. S., Chapter 8 in ref. 2 and Chapter 2 in ref. 6.
51. PELLOW, C., private communication.
52. ZERBI, G. and PISERI, L., *J. Chem. Phys.*, 1968, **49**, 3840.
53. LEADBETTER, A. and LECHNER, R. E. in: *The Plastically Crystalline State*, J. N. Sherwood, Ed., 1979, John Wiley & Sons, New York.
54. ALLEN, G. and HIGGINS, J. S., *Macromolecules*, 1977, **10**, 1006.
55. CONNOR, T. M. and HARTLAND, A., *Polymer*, 1968, **9**, 591.
56. BLEARS, D. J., CONNOR, T. M. and ALLEN, G., *Trans. Far. Soc.*, 1965, **61**, 1097.
57. CRISSMAN, J. M., SAUER, J. A. and WOODWARD, A. E., *J. Polym. Sci. A–2*, 1961, 5075.
58. STEJSKAL, E. O. and GUTOWSKY, H. S., *J. Chem. Phys.*, 1958, **28**, 388.
59. EISENBERG, A. and REICH, S., *J. Chem. Phys.*, 1969, **51**, 5706.
60. MA, K. T., *PhD Thesis*, 1981, Dept. Chem. Eng. and Chem. Tech., Imperial College, London.
61. HIGGINS, J. S. and MA, K. T., to appear in *Polymer*.
62. POWLES, J. G., STRANGE, J. H. and SANDIFORD, D. J. H., *Polymer*, 1963, **4**, 401.
63. CONNOR, T. M. and HARTLAND, A., *Phys. Lett.*, 1966, **23**, 622.
64. LOVELL R. and WINDLE, A. H., *Polymer*, 1981, **22**, 175.
65. TWISTLETON, J. F. and WHITE, J. W., *Polymer*, 1972, **13**, 41.
66. HOLLIDAY, L. and WHITE, J. W., *Pure Appl. Chem.*, 1971, **26**, 545.
67. LYNCH, J. E., SUMMERFIELD, G. C., FELDKAMP, L. A. and KING, J. S., *J. Chem. Phys.*, 1968, **48**, 912.
68. MYERS, W., SUMMERFIELD, G. C. and KING, J. S., *J. Chem. Phys.*, 1966, **44**, 189.
69. PEPY, G. and GRIM, H. in: *Neutron Inelastic Scattering*, 1978, IAEA, Vienna, p. 605.
70. PISERI, L., POWELL, B. M. and DOLLING, G., *J. Chem. Phys.*, 1973, **58**, 158.
71. WHITE, J. W. in: *Dynamics of Solids and Liquids by Neutron Scattering*, S. W. Lovesey and T. Springer, Eds, 1977, Springer Verlag, Berlin.
72. MACONNACHIE, A. and RICHARDS, R. W., *Polymer*, 1978, **19**, 739.
73. ALLEN, G., GHOSH, R., HIGGINS, J. S., COTTON, J. P., FARNOUX, B., JANNINK, G., WEILL, G., *Chem. Phys. Lett.*, 1976, **38**, 577.
74. HIGGINS, J. S., ALLEN, G., GHOSH, R. E., HOWELLS, W. S. and FARNOUX, B., *Chem. Phys. Lett.*, 1977, **49**, 197.
75. AKCASU, A. Z. and HIGGINS, J. S., *J. Polym. Sci., Polym. Phys.*, 1977, **15**, 1745.

76. NICHOLSON, L. K., HIGGINS, J. S. and HAYTER, J. B., *Macromolecules*, 1981, **14**, 836.
77. ALLEGRA, G. and GANNAZZOLI, F., *J. Chem. Phys.*, 1981, **74**, 1310.
78. BRUEGGMAN, B. G., MINNICK, M. G. and SCHRAG, J. L., *Macromolecules*, 1978, **11**, 119.
79. RICHTER, D., HAYTER, J. B., MEZEI, F. and EWEN, B., *Phys. Rev. Lett.*, 1978, **41**, 1484.
80. EWEN, B., RICHTER, D. and LEHNEN, B., *Macromolecules*, 1980, **13**, 876.
81. RICHTER, D., EWEN, B. and HAYTER, J. B., *Phys. Rev. Lett.*, 1980, **45**, 2121.
82. HIGGINS, J. S., NICHOLSON, L. K. and HAYTER, J. B., *Polym. Preprints*, 1981, **22**(1), 86.
83. DE GENNES, P. G., *J. Chem. Phys.*, 1971, **55**, 572.
84. HIGGINS, J. S., GHOSH, R. E., ALLEN, G. and MACONNACHIE, A., *J. Chem. Soc. Faraday II* (in press).
85. HIGGINS, J. S., GHOSH, R. E., HOWELLS, W. S. and ALLEN, G., *J. Chem. Soc. Faraday II*, 1977, **73**, 40.
86. WALSH, D. J., HIGGINS, J. S. and HALL, R. H., *Polymer*, 1979, **20**, 951.
87. HIGGINS, J. S., MA, K. and HALL, R. H., *J. Phys. C. Solid State Physics*, 1981, **14**, 4995.
88. WARNER, M., *J. Phys. C. Solid State Physics*, 1981, **14**, 4985.
89. BENMOUNA, M. and AKCASU, A. Z., *Macromolecules*, 1980, **13**, 409.

Chapter 5

ULTRASONIC CHARACTERISATION OF SOLID POLYMERS

RICHARD A. PETHRICK

Department of Pure and Applied Chemistry, University of Strathclyde, Glasgow, UK

SUMMARY

Ultrasonic propagation has been traditionally used as a high frequency dynamic mechanical test method. In this chapter both the traditional and newer applications of this method are discussed in the context of homo- and hetero-phase polymer systems. Using high frequency sound waves it is possible to sense the increase in modulus associated with fibre formation in drawn polyethylene and polypropylene. The attenuation characteristics of drawn polymers can be explained in terms of a combination of relaxation processes due to the finite time for energy diffusion between spherulites and scattering from both air voids and spherulites. Recent studies of polymers indicate that ultrasonics has uses in the characterisation of polymers other than those traditionally associated with this technique.

1. INTRODUCTION

In an earlier volume in this series the application of ultrasonics to the characterisation of molecular motion in dilute polymer solutions was reviewed.[1] At low frequencies, librational motions coupled with conformational transitions lead to 'normal mode' relaxations, which are dependant upon molecular weight and independant of chemical structure. In contrast, relaxations occurring above 10 MHz are often markedly influenced by chemical structure and can be associated with uncoupled conformational changes involving typically six to eight chemical bonds. A rotational isomeric model of polymer dynamics indicates that the relaxation spectrum of an isolated polymer molecule is composed of two components:[2, 3] a 'normal mode' relaxation involving coherent motions of the whole chain, and segmental motions involving spatial changes of isolated groups of monomers. A more detailed discussion of this topic will be published elsewhere.[2, 3] Ultrasonic measurements of solid polymers may be considered to be the high frequency analogue of low frequency dynamic mechanical studies and therefore the relaxations are usually concerned with conformational changes of elements of the chain of varying size. However, recent research has indicated that the wavelength dependent characteristics of sonic propagation may provide information of morphologically related phenomena and it is this area of study which will form the main body of this review.

2. TECHNIQUES FOR SOUND PROPAGATION MEASUREMENTS IN SOLID POLYMERS

Sound waves can propagate in an unbound isotropic solid in two ways: either as a wave which generates vibrations along its direction of propagation—*longitudinal* mode motion, or as a wave which generates vibrations perpendicular to its direction of propagation—*shear* modes.

2.1. Immersion Method

The simplest and most widely used method of measuring the velocity (v) and attenuation (α) of sound in the frequency range 1–1000 MHz,[4–6] is the immersion method. A typical system (Fig. 1) consists of a high frequency, high voltage electrical pulse generator, some means of amplification of weak signals and a display oscilloscope. The electrical pulses are first applied to a piezoelectric transducer whereupon they are converted into pressure waves of an appropriate frequency. The high voltage oscillator is pulsed and a typical wave packet will have duration 1 ms, allowing sufficient time for pure simple harmonic oscillation to be established. The propagating sound wave is detected by a second transducer and converted into an electrical signal which is subsequently amplified and displayed. A typical immersion cell (Fig. 2) allows the rotation of a clamped sample relative to the axis of the sound beam. Comparison of the intensities of the

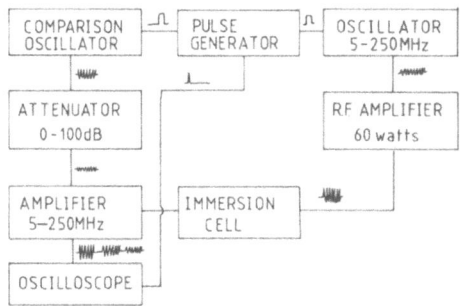

FIG. 1.   Typical ultrasonic immersion apparatus. The type of piezoelectric transducer used depends on the frequency range to be investigated. For frequencies of
100 MHz and below an X-cut quartz crystal is employed, whereas for measurements
above 100 MHz a CdS or lithium niobate crystal is favoured.

received pulse with and without the sample allow the longitudinal attenuation to be determined. For these latter measurements the sample is
aligned perpendicular to the direction of propagation of the sound wave.
    Rotation of the sample in the sound beam leads to the observation of two
sets of distinct maxima in the attenuation (Fig. 3). The first set corresponds

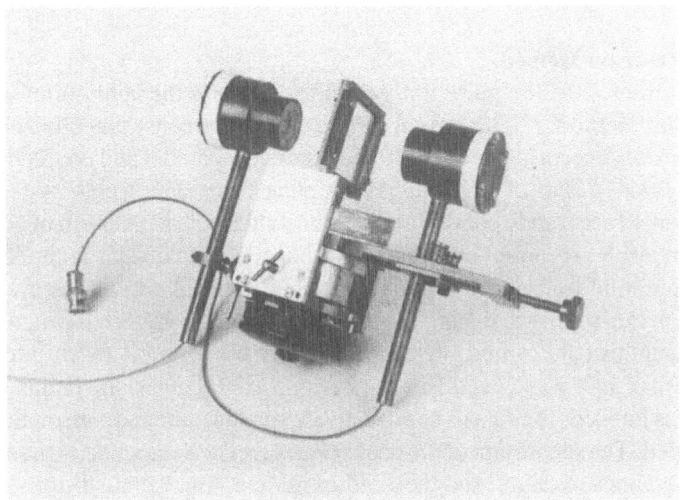

FIG. 2.   Typical immersion cell. The system is usually placed in a thermostatted
bath containing water, alcohol or silicone fluid as immersion liquid.

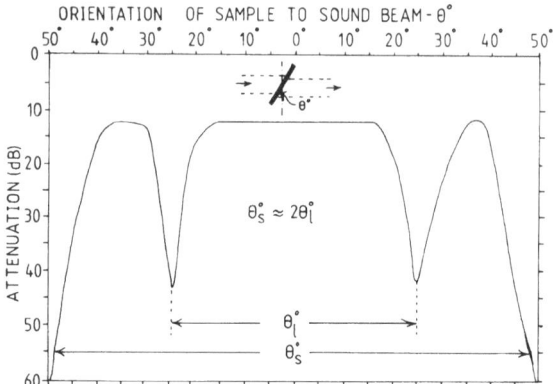

FIG. 3. Plot of acoustic attenuation versus angle for polyethylene. The closest spaced set of maxima in the attenuation correspond to the condition for critical refraction as defined by Snell's law and allow determination of the longitudinal propagation velocity. The second set of maxima correspond to the shear wave propagation and occur at roughly twice the angle for longitudinal wave propagation.

to the critical refraction condition for the longitudinal wave and the second set to that for shear wave propagation.

2.2. Resonant Method

An alternative to the pulse immersion technique is the continuous wave resonant method.[7, 8] The typical system (Fig. 4), is constructed from two quartz transducers aligned so as to be accurately parallel and coupled to a parallel sided slab of polymer via coupling films. The whole system is maintained accurately parallel and thermostatted with a precision of better than $\pm 0 \cdot 01$ K. A continuous wave signal of variable frequency is applied to the excitation transducer and a synchronously tuned detector is used to amplify the received signal. An electrical signal is detected when the wavelength of the sound wave in the cavity corresponds to an integral number of half wavelengths of the cavity's axial dimension. If this condition is fulfilled, the signals constructively add together and a resonance is observed. The separation of the peaks (Fig. 4) reflects the velocity of sound in the composite cavity, and their widths represent the attenuation.

The above methods allow determination of the attenuation of sound (α) with a precision of $\pm 2\%$ and the velocity (v) with a precision of between 2% (immersion method) and $0 \cdot 1\%$ (resonant method).

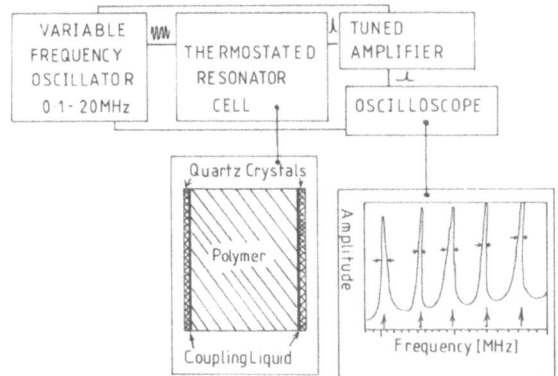

FIG. 4. Typical resonator apparatus. An electrical signal is detected when the separation between the excitation and detection crystals corresponds to an integral number of half wavelengths. The width of the resonance is a function of the coupling efficiency of the crystals to the polymer and also of the intrinsic loss in the polymer.

3. THEORY OF SOUND PROPAGATION IN POLYMERS

In the limit of small amplitude deformations Hooke's law is obeyed and the strain ϵ is a linear function of the stress:

$$\sigma = E\epsilon \tag{1}$$

where E is the Young's modulus of the system. In the case of a viscoelastic medium, the modulus is a mathematically complex quantity $E^* = E' + iE''$, where E' is the real and E'' the imaginary part of the modulus. The bulk modulus, K_s^*, has the form:

$$K_s^* = K_s' + iK_s'' = \frac{E^*}{3} \frac{1}{1 - 2\mu^*} \tag{2}$$

where μ^* is the complex Poisson ratio. Equation (2) is the starting equation for the discussion of acoustic propagation in solids. At high frequency, the wavelength of the sound wave is much less than the lateral dimensions of the sample, thence the velocity is that of a purely longitudinal wave (v_1) which has the form:

$$v_1 = \sqrt{\frac{K' + 4/3G'}{\rho}} = \sqrt{\frac{E'(1-\mu')}{\rho(1+\mu')(1-2\mu')}} \tag{3}$$

where ρ is the density of the medium and G' is the corresponding shear modulus. It is convenient to define an equivalent dynamic longitudinal modulus (L') written as:

$$v_1 = \sqrt{L'/\rho} \tag{4}$$

where $L^* = L' + iL''$, which is also equal to $K_s^* + 4/3G^*$. When the wave propagation is constrained to thin strips ($\lambda \leq b$, the lateral dimensions of a body) eqn (4) becomes:

$$v_1 = \sqrt{E'/\rho} \tag{5}$$

In the case of oscillations occurring normal to the direction of propagation of the equivalent transverse or shear wave velocity:

$$v_t = \sqrt{G'/\rho} \tag{6}$$

The above equations apply only if the attenuation (α) in the solid is small: $\alpha\lambda/2\pi \leq 1$. In polymers, it is necessary to allow for the effects of dispersion in the calculation of the equivalent elasticity modulus $M^* = M' + iM''$ given by:

$$M' = \frac{\rho v^2 [1 - (\alpha\lambda/2\pi)]^2}{[1 + (\alpha\lambda/2\pi)^2]^2} \tag{7}$$

$$M'' = \frac{2\rho v^2 (\alpha\lambda/2\pi)}{[1 + (\alpha\lambda/2\pi)^2]^2} \tag{8}$$

If v is replaced by v_1 the elastic modulus M^* becomes the longitudinal modulus, L, and alternatively if v_s is used then the shear modulus G is obtained.

The attenuation coefficient α describes the variation with distance of the sound intensity and is described by:

$$I_x = I_o e^{-\alpha x} \tag{9}$$

where I_x and I_o are respectively the intensities of the sound wave at a point x and at the origin.

4. MOLECULAR INTERPRETATION OF THE ACOUSTIC PARAMETERS

The value of the dynamic modulus will change with temperature; at low temperatures where molecular motion in the solid is frozen, the modulus will have a value of 10^{10} dyn/cm^2. Increasing the temperature will lead to the formation of a rubbery phase which will have a correspondingly lower value of the modulus—10^6–10^7 dyn/cm^2. A decrease in modulus is associated with the onset of molecular motion in the solid state. The dynamic modulus of elasticity and the sound velocity are determined by the bond energy of the atoms forming the main backbone of the polymer (intramolecular) and the interaction between elements on adjacent polymer chains (intermolecular). In theory, the velocity of sound along the backbone may have a value of the order of 8–12×10^3 m/s and is principally determined by intramolecular interactions. The shear velocity yields values of the order of $1 \cdot 2$–$1 \cdot 5 \times 10^3$ m/s and reflects the strength of intermolecular van der Waals interactions.

Associated with the occurrence of active conformational change in the solid is the observation of a peak in the loss modulus. When the frequency of perturbation is comparable to $1/\tau$, where τ is the relaxation time, a maximum is observed in the loss modulus. For a simple thermally activated process:

$$\tau^{-1} = A \exp(-\Delta E/RT) \qquad (10)$$

where ΔE and A are respectively the activation energy and pre-exponential factor reflecting the mechanism of conformational change detected. At low frequencies, the system can deform and conformational change occur within the period of oscillation of the sound wave. The modulus at low frequencies has a value appropriate to that of a rubbery phase. Alternatively, at high frequencies the perturbation occurs much faster than the rate of molecular reorientation and the value of the modulus approximates to that of the glassy phase. In between these extremes the modulus is observed to drop in a sigmoidal fashion with increasing temperature and the energy loss exhibits a maximum. The relative ease of rotation is reflected in the position on the frequency axis of the loss peak. The activation energy can be obtained from an investigation of the temperature dependence of the relaxation frequency. The magnitude of the activation energy reflects the nature of the molecular process being probed and can be classified as follows:

(i) alpha process—associated with the glass transition temperature T_g and involving relatively large scale motion of main backbone elements exhibiting activation energies between 80–800 kJ/mol.

(ii) beta process—associated with the rotational isomerism of side chain elements—activation energies in the range 60–160 kJ/mol.

(iii) gamma processes—also associated with rotation of groups attached
to the main backbone or of side chains—activation energies in the
range 28–80 kJ/mol.

(iv) delta processes—associated with rotational motion of groups
attached to side chains—activation energies in the range 16–40
kJ/mol.

The alpha process is a co-operative relaxation and corresponds to a
combination of segmental motion and certain of the higher normal modes.
The selection of the latter will depend on the nature of the polymer–
polymer interactions occurring in the solid and will usually correspond to
motions involving six to ten monomer units.

The above summary represents the behaviour found in amorphous
polymers. If the polymer is capable of forming a partially crystalline solid,
a multiplicity of relaxations will be observed corresponding to motions in
the crystalline and amorphous phases.

5. REVIEW OF STUDIES OF THE HIGH FREQUENCY
RELAXATION BEHAVIOUR OF SOLID POLYMERS

Reviews of ultrasonic relaxation in solid polymers have been published
elsewhere.[9–11] The interpretation of the data has to a large extent neglected
the morphology of the solid. Recent advances in the methods available for
the characterisation of solid polymers have allowed much of this earlier
data to be reinterpreted.[12, 13] In this next section, examples of the influence
of morphology on the form of the ultrasonic relaxation observed in
polymers will be discussed.

5.1. Polyvinylchloride (—CH$_2$—CHCl—)$_n$

A comparatively simple and relatively rigid polar polymer, it exhibits a
range of relaxation features.[14–25] A close examination of the velocity
dependence in the region of the T_g indicates the existence of a double
transition, attributed to the occurrence of super-molecular organisation in
the solid. It has been postulated that the polymer chains in PVC may
undergo specific interactions leading to regions of higher density. Crystallinity is inhibited in this polymer by a lack of a regular stereochemistry of
the main backbone and the possibility of branched chain structures occurring at irregular intervals along the polymer backbone. The postulate of
specific interactions and super-molecular order is supported by the observation that annealing leads to an increase in the density associated with

growth of the crystalline regions. Clearly, polymer intramolecular interactions, even in so-called amorphous polymers can play a very important
role in the determination of the relaxation spectrum of a particular macromolecule.

5.2. Polystyrene ($-CH_2CHPh-$)$_n$

Polystyrene is typical of many polymers, exhibiting a multiple relaxation
spectrum and appearing from X-ray analysis to be completely amorphous.[24-29] Dielectric, NMR and low frequency dynamic mechanical
studies have indicated that the transition detected acoustically at 193 K
can be ascribed to the onset of oscillatory motion of the phenyl group.
Transitions observed at 383 and 373 K are attributed to the glass transition
and this assignment has been confirmed by dilatometry measurements.
The observation of two transitions leads to the proposition that clustering
occurs in the solid phase. The lower temperature transition would therefore correspond to the onset of collective motions of polymer chains in the
amorphous regions. The low temperature dependence of the modulus
between 373 and 383 K is believed to be a consequence of the reinforcing
action of the clustered regions. Recent ultrasonic measurements of
polymer solutions[30] have indicated that at 340 K a marked increase in the
relaxation amplitude occurs associated with a helix–coil type of transition.
Conformational transitions associated with certain stereoregular
sequences do not become active until high temperatures and in the
presence of the stronger intermolecular forces found in the solid may be
expected to be shifted to higher temperatures. It is therefore possible that
part of the reason for multiple relaxation in the region of T_g may be related
to the distribution of stereoregular forms found in the 'atactic' polymer.

Investigation of the temperature dependence of the velocity of sound
above 419 K indicates that it becomes essentially independent of temperature corresponding to a plateau modulus condition. Analysis of the data
using the kinetic theory of rubber elasticity yields the relation:[31]

$$E_o = 3\rho RT/M_c \qquad (11)$$

where ρ is the density and M_c is the molecular weight of the polymer chain
between entanglement points. Using the relation $v^2 = E_o/\rho$, eqn (11)
becomes:

$$v^2 = 3RT/M_c \qquad (12)$$

Analysis of the data above 423 K using eqn (12) indicates that M_c has a
value of approximately 4580. This implies that there are on average 44
monomer units between each entanglement.

5.3. Poly(alkylmethacrylate)s

Poly(methyl methacrylate) (PMMA) (Structure I) has been investigated extensively.[32, 33] The T_g process in the atactic polymer appears in the acoustic spectrum at 390 K. The ester group motion occurs at approximately 291 K and rotation of the methoxy and methyl groups have been

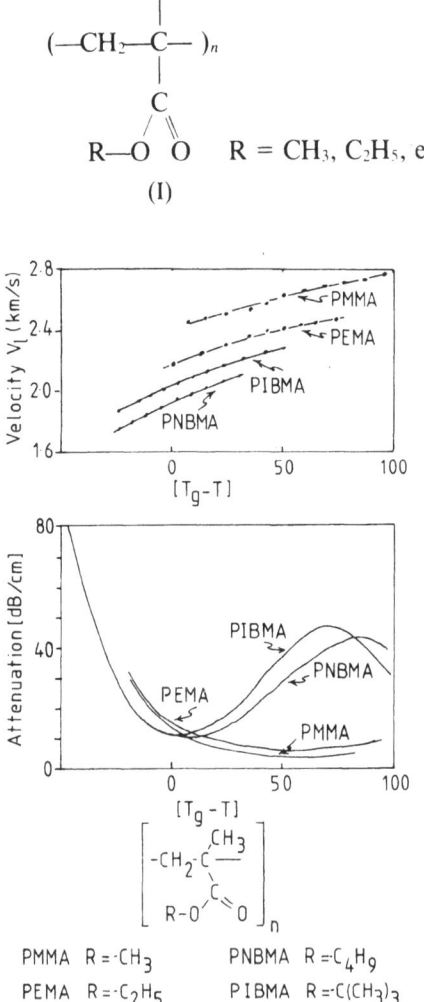

FIG. 5.   Ultrasonic attenuation in poly(alkylmethacrylate)s.

assigned respectively to transitions at 233 and 166 K. This assignment has recently been supported by neutron scattering studies on deuterated PMMA.[34] Partial substitution of the backbone and side chain hydrogens allows the motions of the methyl group to be unambiguously identified. In the case of alkyl substituted methacrylate polymers, an additional transition is detected ultrasonically [35] (Fig. 5), associated with the onset of motion of the hydrocarbon chain. Similar conclusions have been obtained from the analysis of NMR and dynamic mechanical data.[36–38] The motion of

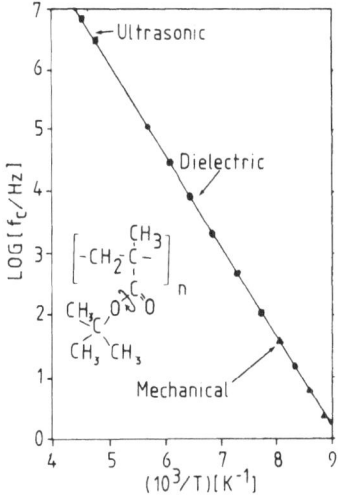

FIG. 6. Correlation diagram for butyl side chain motion in poly(butyl methacrylate), ultrasonic data from reference 35, dielectric data from reference 42 and mechanical data from reference 43.

the side chain in the case of the butyl group exhibits a simple thermally activated behaviour (Fig. 6), in contrast to the relaxation of the backbone which is free volume controlled. The assignment of the transition at 291 K in PMMA is a little more difficult. It has been ascribed to 'retarded rotation of the methoxy carbonyl group of the side chain'. It is not, however, clear why we should apparently have two relaxations assigned to the same motion. An alternative explanation may be found by consideration of the tacticity of PMMA. The atactic polymer will be composed of syndio, hetero and isotactic sequences. The T_gs of these sequences are respectively 388 (syndio) and 318 K (iso). It is possible that the incorporation of the iso sequence into an atactic polymer will increase its mobility and lower the value of the effective T_g. It is therefore possible that the transition at 291 K may in part arise from relaxation of the isotactic sequences coupled with the motion of the side chain groups.

As in the case of polystyrene, a plateau is observed in the velocity/modulus above 431 K. Application of eqn (12) indicates that M_c is equal to 1985 yielding a mean network dimension for PMMA of 20 monomers between crosslinks. This value is appreciably lower than that observed in the case of polystyrene and probably reflects the effects of specific interactions between stereoregular sequences in the polymer.

5.4. Poly(vinyl acetate)
In contrast to PMMA, poly(vinyl acetate) (Structure II) exhibits four relaxations in the acoustic spectrum:[39-41] 153, 195, 288 and 306 K. The lowest temperature transition may be ascribed to methyl group rotation and this is in agreement with neutron scattering observations. Dilatometric

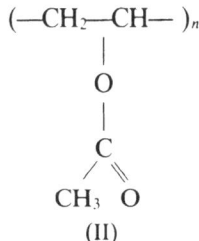

$$(-CH_2-CH-)_n$$
$$|$$
$$O$$
$$|$$
$$C$$
$$CH_3 \quad O$$
(II)

studies indicate the existence at 306 and 313 K of two transitions connected with the onset of molecular motion of the backbone in amorphous and clustered regions.[42, 43] This is in agreement with dielectric and NMR studies of this polymer. The transition at 193 K is assigned to the rotational isomerism of the side chain groups.

A plateau corresponding to rubbery behaviour is observed above 310 K. Application of eqn (12) yields a value of M_c equal to 4680 which indicates that on average there are 55 monomer units between entanglements.

5.5. Polycarbonate, Poly(ether sulphone) and Polysulphone
These polymers possess a *para*-linked chain extended diphenyl methane structure (Fig. 7). They are the so-called 'engineering polymers' having high melting points and impact strengths. The origin of the high impact strength is still a matter of some discussion; facile motion of the molecular backbone below T_g has been considered to be important as has relative motion of the polymer chains.[44-46] Rotation of the backbone has been investigated using

ultrasonic and dielectric relaxation techniques.[47–49] All of these polymers exhibit a broad relaxation in the frequency range 100 to 10^6 Hz at room temperature. A comparison of the temperature dependence of the dielectric and low frequency mechanical measurements of polycarbonate have shown two major relaxation processes: the alpha process with activation energy 450 kJ/mol associated with large scale motion of the polymer chain; the beta process with activation energy of 45 kJ/mol ascribed to more local reorientational motions of the phenyl ring structures. An additional beta+ process has been detected in certain

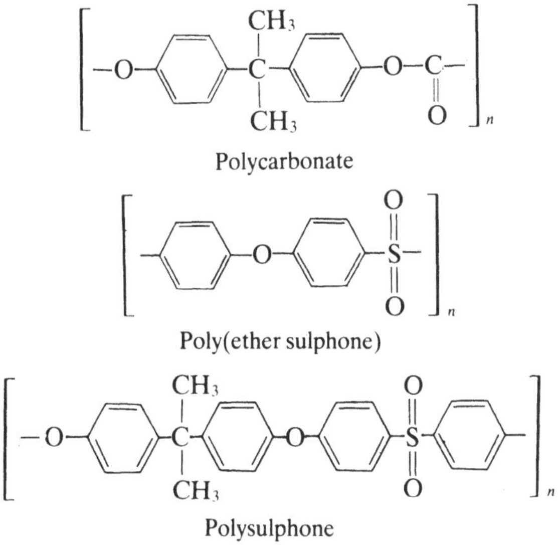

Polycarbonate

Poly(ether sulphone)

Polysulphone

FIG. 7.   Structures of polycarbonate, poly(ether sulphone) and polysulphone.

samples and is attributed to either changes in morphology or relaxation of strained regions within the sample. The activation energies for the beta process in poly(ether sulphone) and polysulphone are respectively $13 \cdot 1 \pm 2$ kJ/mol and $16 \cdot 6 \pm 2$ kJ/mol.[50] The activation energy plot for polycarbonate is shown in Fig. 8.

The temperature dependence of the ultrasonic velocity (Fig. 9), normalised to the glass transition temperature, indicates that both polycarbonate and poly(ether sulphone) possess similar molecular relaxation characteristics. In contrast, polysulphone either has a much higher sound velocity (and modulus) in the rubbery state or more likely undergoes another relaxation (with associated decrement in the sound velocity) between

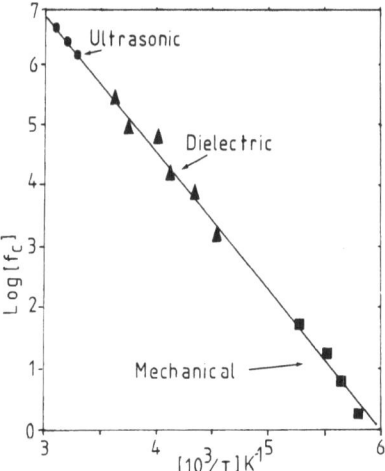

FIG. 8. Correlation plot for the beta relaxation in polycarbonate. Ultrasonic data from reference 50, dielectric data from reference 48 and mechanical data from reference 47.

363 K and the glass transition. This is probably a consequence of partially crystalline structures existing in the solid at room temperature.

The high impact strength of these materials is ascribed to the beta relaxation: it is assumed that the local stress concentration generated by the impact may be dissipated by the facile molecular motion avoiding the rupture of the polymer backbone bonds. Polycarbonate is exceptional in

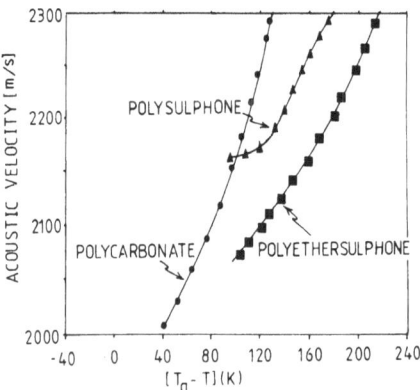

FIG. 9. Ultrasonic velocity normalised to the glass transition temperature for polycarbonate, poly(ether sulphone) and polysulphone.

that it possesses a free volume which is 1·5 times greater than that observed for the majority of polymers. As a consequence the possibility of motion of one chain relative to another must be included in the discussion of the mechanism of the response to impact.

Measurements of the sound velocity in the rubbery plateau region yield values of M_c between 1680 and 2300, implying that the number of chain elements between entanglement is very low compared with other polymers and is of the order of 6–9 units.

5.6. Polyolefin Rubbers

A large number of investigations have been reported on natural and synthetic rubber.[51-56] Natural or hevea rubber is obtained from *Hevea brasiliensis*. It has the chemical structure of poly(*cis*-1,4-isoprene), Fig. 10.

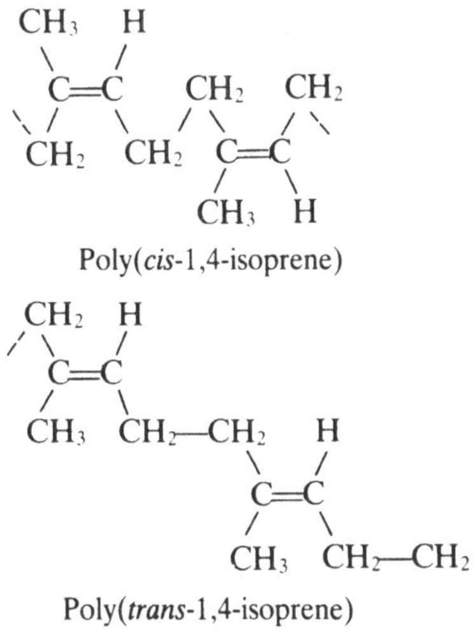

Poly(*cis*-1,4-isoprene)

Poly(*trans*-1,4-isoprene)

FIG. 10.   Chemical structures for poly(*cis*-1,4-isoprene) and poly(*trans*-1,4-isoprene).

The presence of the C=C bond imparts a conformational rigidity to the backbone structure. Gutta-percha is also naturally occurring, however it has the *trans* rather than *cis* structure (Fig. 10). This simple change in the backbone structure leads to very different physical properties: hevea is amorphous and rubbery at room temperature, flows at 333 K and crystallises on cooling below 273 K or on stretching. In contrast, gutta percha is a highly crystalline material at room temperature. The *cis* polymer in the crystalline state has a unit cell containing two right handed and two left handed cis-4-isoprene units.

Synthetic polymers are obtained by use of Ziegler–Natta catalysis which leads to a poly(cis-1,4-isoprene) structure. These materials are usually strengthened by vulcanisation with either sulphur or a peroxide treatment. This process involves the generation of cross-linkages which are either sulphur or oxygen bridges.

Acoustic studies indicate that rubbers produce a large attenuation of the sound wave in the frequency range 40 kHz to 10 MHz. The magnitude, rate of change with frequency and position of the loss are all sensitive to the effects of change in stereochemistry of the backbone. Hevea rubber has the highest loss and gutta percha the lowest. Commercial rubbers lie between these extremes. Fillers such as carbon, iron oxide, china clay, magnesium oxide and lithage have a significant effect on the loss and velocity of sound propagation.[57] The fillers appear to slow down the motion of the polymer chains and increase the modulus.

5.7. Polytetrafluoroethylene ($-CF_2-CF_2-$)$_n$

This is a non-polar, linear and highly crystalline polymer. Commercial PTFE is usually the product of sintering and has its crystallinity reduced by the introduction of branched chains. Crystalline (93–98%) PTFE has a melting point of 620 K, and exhibits a helical structure which undergoes a unit cell transition at 292 and 303 K.[58–60] Below 292 K the conformation can be described by a helix containing 13 CF_2 units per 180° twist. The unit cell is triclinic and essentially perfect three-dimensional order is observed in the polymer. Above 292 K the helix expands to 15 CF_2 groups per twist of 180° and the unit cell becomes hexagonal. Between 292 and 303 the packing of the molecules on the hexagonal lattice is disordered due to small displacements arising from angular oscillations of segments about their long axis. Above 303 K the preferred crystallographic direction is lost and segmental oscillations lead to random angular orientation on the lattice. The crystal–disorder transitions at 292 and 303 K have been studied by X-ray diffraction,[58] densiometry[59] and specific heat studies.[60]

Acoustic studies of crystalline PTFE report discontinuities in the absorption–temperature plots at 292 and 303 K and broader transitions at 173 and 393 K, Fig. 11. The very sharp peaks clearly correspond to the crystalline order–disorder transitions discussed above.[61, 62] The peak at 393 K is sensitive to the degree of crystallinity of the sample, increasing in amplitude with increasing degree of crystallinity and has been assigned to torsional oscillation of chain segments around the chain axis. Studies of the ultrasonic relaxation in solution[63] have indicated that the conformational relaxation associated with an exchange between the non-degenerate *trans* structures should occur in the region observed for the conformational transition in the solid and explain the above behaviour.

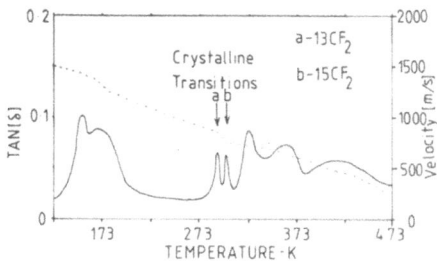

FIG. 11.   Ultrasonic absorption as a function of temperature for polytetrafluoro-ethylene.

5.8. Poly(vinylidene fluoride): ($-CH_2-CHF-$)$_n$ Polytrifluoroethylene: ($-CHF-CF_2-$)$_n$, Polychlorotrifluoroethylene: ($-CFCl-CF_2-$)$_n$

These polymers, like PTFE, exhibit relaxation features which reflect the existence of crystalline structures in the solid state. It has long been recognised that the substitution of fluorine atoms for hydrogen in poly-ethylene leads to a reduction in the intermolecular attractive forces.[64] However, nearly all of the ultrasonic studies on fluorinated polyethylene have concentrated on the relaxation region and have not attempted to quantify the magnitude of these changes. Extrapolation of the ultrasonic velocity data from high temperatures indicates that the low temperature intermolecular potentials of these polymers are in fact significantly different.[65] Combination of the velocity data with thermal observations allows an effective three-dimensional intermolecular force field to be defined.[66-70] Using the model of Tarasov it is possible to confirm the existence of low frequency 'vibrational' modes which contribute to the low temperature specific heat of these polymers. As we shall see later, this phenomenon may in part be associated with the time dependence of thermal diffusion between crystalline domains.

5.9. Polyamides—Nylon

Nylon is the condensation product of a diamine and a dicarboxylic acid:

$$[-NH-(CH_2)_x-NH-CO-(CH_2)_y-CO-]_n$$

For nylon 66 $x = 6$ and $y = 6$. The properties of nylon are dominated by the formation of the intermolecular N—H---O=C bonds. Crystallisation from solution leads to the formation of lamellar structures of thickness between 500 and 1000 nm. The bulk polymer usually forms spherulites orientated so that the hydrogen bonds are along the radius. In normal bulk polymer, the crystallinity can vary between 20 and 55%.

Acoustic relaxation peaks observed at room temperature[71] are sensitive to water content and have been ascribed to relaxation of the amide group and its solvation sheath.[72-76] A similar feature is observed in the dynamic mechanical spectrum at 238 K and leads to an activation energy of 10 kJ/mol for this process.

The longitudinal velocity exhibits a change in slope at approximately 348 K and this is associated with the onset of motion in the crystalline regions of the polymer.[73, 74] The shear velocity shows a similar trend, the change in slope being less dramatic than in the longitudinal case.

5.10. Polyethylene (—CH₂—)ₙ

This polymer has the simplest of structures; however, it exhibits one of the most complex of relaxation spectra. It is one of the few polymers to have been investigated by ultrasonics in the region of its melt temperature—T_m. The precise value of T_m depends on the crystallinity but usually lies in the range 389 to 403 K.[77-81] The denser, more highly crystalline samples have the higher melt temperatures. The melt temperature is marked acoustically by a sharp decrease in both the longitudinal and transverse velocities.

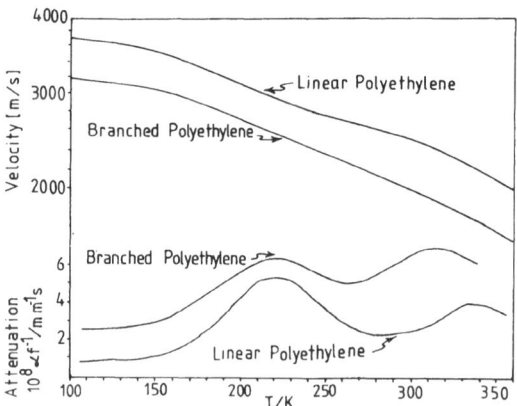

FIG. 12.   Ultrasonic absorption as a function of temperature for polyethylene.

Observations at low frequency (50 kHz) have indicated the existence of a premelt increase in the attenuation which has been ascribed to the onset of molecular motion in areas of the crystalline phase.[82, 83] A marked drop in the acoustic velocity in the temperature range 213 to 223 K is associated with the glass transition of the amorphous phase. Increase in the branched chain content has a significant effect on the loss in the room temperature range and only a minor effect on the lower temperature process, Fig. 12. Clearly the higher temperature process is associated with the motion of the pendant groups; however the insensitivity of the lower temperature process to the effects of change in crystallinity indicates that it is not simply

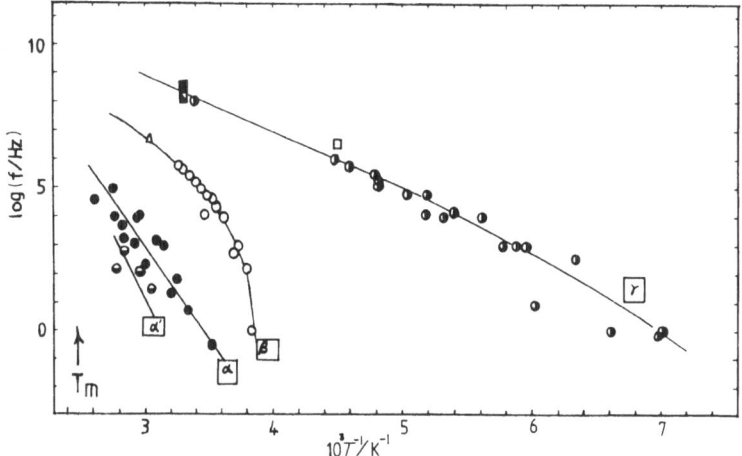

FIG. 13.   Correlation diagram for polyethylene.

the result of motion of chains between the crystalline blocks. A correlation plot of the relaxations, with tentative assignments of processes is presented in Fig. 13. Spin probe, dielectric and low frequency dynamic mechanical studies on linear and gamma irradiated material have recently helped to elucidate the nature of the relaxation processes in this polymer and these data are included in the correlation plot.

5.11. Polypropylene ($-CH_2-CHCH_3-)_n$

As in polyethylene, the relaxation of polypropylene is complicated by the existence of paracrystalline regions. A well defined loss process has been observed using ultrasonics in the frequency range 10 to 300 kHz at room temperature and has been assigned to large scale motion of the amorphous regions of the polymer.[84, 85] Neutron scattering data on partially deuterated polymers have provided an unambiguous assignment of the rotational relaxation of the methyl group. As in the case of polyethylene, the relaxation spectrum is sensitive to the effects of thermal annealing.

It is clear from the above review, that the acoustic relaxation of many so-called amorphous polymers are far from simple and that many of the 'additional' features arise as a consequence of order in the solid phase. It is not always clear whether this order is inter- or intra-chain and data from techniques such as electron microscopy, X-ray and neutron diffraction can assist greatly in the assignment of the origins of the effects observed.

6. WAVELENGTH AND VECTORIAL PROPERTIES OF SOUND PROPAGATION

6.1. Drawn, Anisotropic Polymers

Many polymers, polyethylene being a classic example, can form crystalline ordered regions in their solid state.[86] Mechanical work in terms of drawing or compression can achieve alignment of the axes of these crystalline regions with a consequent increase in the modulus of the material in the direction of draw and consequent changes in the relaxation spectrum. The morphology of the crystalline state of polyethylene has been extensively investigated. Using X-ray diffraction techniques it is possible to indicate the alignment of a polymer in terms of its 'pole' diagram.[87, 88] Similar data on the orientation of the crystalline phase in the polymer can be obtained from acoustic experiments. In the typical immersion experiment the velocity is determined perpendicular to the axis of rotation of the sample holder. If the direction of alignment is varied it is possible to explore the various elastic components associated with the anisotropic nature of the crystalline phase. Examples of typical 'acoustic pole' diagrams are presented in Fig. 14. Moseley has proposed a semi-empirical treatment of the propagation equations to describe such an experiment.[89] He considered two different cases: (i) a series addition of force constants and (ii) a parallel addition of force constants. Only the first case appears to lead to valid results and can be described by an equation of the form:

$$\frac{1}{v^2} = \frac{1 - \langle \cos^2\theta \rangle}{v_\perp^2} + \frac{\langle \cos^2\theta \rangle}{v_\parallel^2} = \frac{\langle \sin^2\theta \rangle}{v_\perp^2} + \frac{\cos^2\theta}{v_\parallel^2} \qquad (13)$$

where v, $v_\perp$ and $v_\parallel$ are respectively the ultrasonic velocities for samples

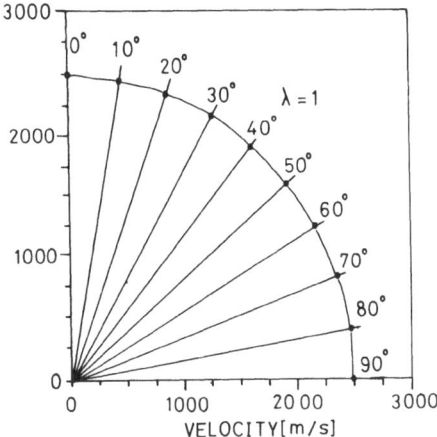

FIG. 14. Acoustic pole diagram for isotropic and drawn polyethylene samples. The diagram is obtained by measuring the velocity at various angles of orientation draw of the sample to the axis of rotation. For an unorientated sample the plot will yield a semicircle, for orientated materials this will be distorted into an elliptical form, the ratio of the major and minor axes reflecting the degree of orientation of the polymer.

having an average orientation angle θ, 90° (perpendicular orientation) and 0° (parallel orientation), respectively. Equation (13) assumes that only the second moment ($\langle \cos^2\theta \rangle$) of the orientation distribution is influential in determining the velocity. Ward[90] has however questioned the assumption and starting with the model that the solid consists of aggregates of structural units whose elastic constants are identical to those of the highly orientated material, obtains the relation:

$$\frac{1}{E} = \frac{\langle \sin^4\theta \rangle}{E_\perp} + \frac{\langle \cos^4\theta \rangle}{E_\parallel} + \langle \sin^2\theta \cos^2\theta \rangle \left[\frac{1}{G} - \frac{2\gamma}{E} \right] \qquad (14)$$

where G refers to the torsional modulus of the perfectly orientated material and γ is its Poisson's ratio. Equation (13) can be compared with eqn (14) using the relation $E = \rho v^2$. Clearly E is not only a function of the second moment of the orientation distribution, but is also dependent on the fourth moment as well. Since γ is of the order of 0·5, the term in $\gamma/E_\parallel$ will be small as will $1/E_\parallel$ relative to $1/G$ or $1/E_\perp$. In this case eqn (14) reduces to

$$\frac{1}{E} = \frac{\langle \sin^4\theta \rangle}{E_\perp} + \frac{\langle \sin^2\theta \cos^2\theta \rangle}{G} \qquad (15)$$

If the assumption is made that $E_\perp \approx G$ which is not unreasonable for moderate orientation, then eqn (15) becomes:

$$\frac{1}{E} = \frac{1 - \langle \cos^2 \theta \rangle}{v_\perp^2} \tag{16}$$

This is identical with the Moseley equation (13), if the second term is negligible. The principle limitation of this method is that it measures the average orientation of the total sample. It has however been found that a good correlation exists between the orientation data derived from ultrasonic measurements and those obtained from birefringence measurements.[91]

Data obtained for polyethylene and polypropylene are summarised[92, 93] in Table 1. The high value of the modulus observed in the draw direction is consistent with sound waves sensing distortion of the polymer chain (intramolecular bond stretching) while the perpendicular propagation senses the

TABLE 1

ELASTIC STIFFNESS CONSTANTS FOR DRAWN POLYMERS

Polymer	Conditions	C_{11} (GPa)	C_{22} (GPa)	C_{33} (GPa)
Polyethylene	Undrawn	3·9	3·9	3·9
	Drawn × 2	4·3	4·3	4·8
	Drawn × 3·5	4·1	4·2	5·1
	Drawn × 8	3·9	3·7	17·9
	Drawn × 11	4·1	4·2	36
	Drawn × 18	4·3	4·4	53
Polypropylene	Undrawn	5·75	5·75	5·75
	Drawn × 1·7	4·7	4·7	6·28
	Drawn × 3·3	3·19	3·4	8·98
	Drawn × 5·5	3·2	3·5	11·3
	Drawn × 10·5	4·16	4·3	15·0
Poly(ethylene terephthalate)	Undrawn 5·8		—	5·8
	Drawn × 2	5·7	—	7·5
	Drawn × 3	5·7	—	10
	Drawn × 4	5·7	—	12·5

The stiffness constants are defined in ref. 13. C_{33} approximates to the force constant in the fibre direction and C_{11} and C_{22} to the constants perpendicular to the draw direction.

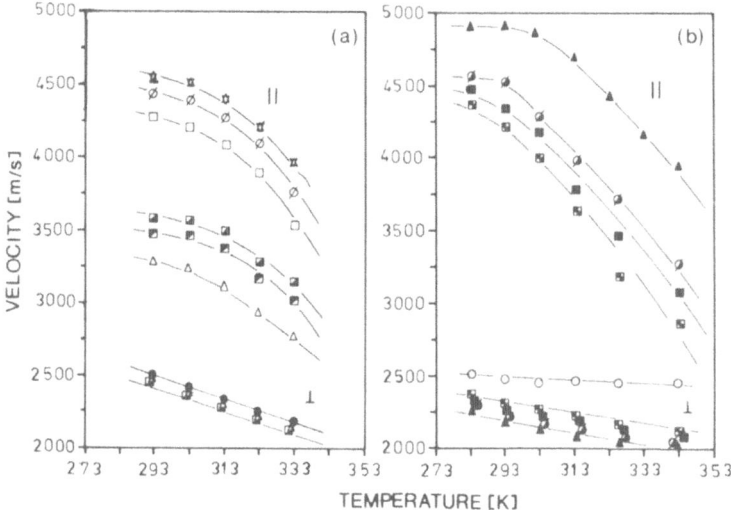

FIG. 15. Temperature dependence of the ultrasonic velocity in (a) polyethylene, and (b) polypropylene. Polyethylene: ●, undrawn polymer; △, $\lambda = 3$; ◪, $\lambda = 4$; ◩, $\lambda = 5$; □, $\lambda = 7$; ϕ, $\lambda = 10$; and $\Diamond$, $\lambda = 12$. Polypropylene: ○, undrawn polymer; ◪, $\lambda = 2{\cdot}6$; ■, $\lambda = 6$; ϕ, $\lambda = 9$; ▲, $\lambda = 12$. The symbols ‖ and ⊥ indicate that the respective curves correspond to the components parallel and perpendicular to the draw direction. Both of the above curves are sensitive to the molecular weight, molecular weight distribution and drawing temperature used in preparing the samples.

modulus transverse to the fibre axis. The latter is determined by inter-molecular bonds and is primarily determined by van der Waals interactions between the fibres. The temperature dependence of the velocity both in polyethylene[94] and polypropylene[95] (Fig. 15), reflects the nature of the molecular motions possible in these samples. A change in slope is indicative of the onset of molecular motion in the sample. Change in the shape of the curve with draw ratio is indicative of changes in the spectrum of relaxation motions which influence the acoustic propagation. In the case of polypropylene, there appears to be a shift of the relaxation spectrum on drawing as indicated by an elbow in the high draw ratio samples. Investigation of the pole diagram of an intermediate draw ratio polyethylene indicates that there is a significant component at 45° to the draw direction.[94] It has been reported from X-ray measurements that in the initial stages of draw the crystalline regions can become orientated at 45° to the draw direction; the ultrasonic observations are in agreement with X-ray data.

6.2. Phase Separation

Samuels[96] has extended the Moseley treatment to the consideration of two-phase systems—specifically semi-crystalline polymers. This theory leads to an equation of the form

$$3/2(\Delta E)^{-1} = \frac{(1-\beta)f_1}{E_1} + \frac{\beta f_2}{E_2} \qquad (17)$$

where β is the volume fraction of component 2 while f_1 and f_2 are, respectively, the Hermans' orientation functions for components 1 and 2 and E_1 and E_2 are the moduli values for the perfectly orientated components and

$$(\Delta E^{-1}) = \left[\frac{1}{E_u} - \frac{1}{E} \right] \qquad (18)$$

where E_u and E are the respective moduli for an unorientated sample and the sample being measured. If we consider the case of a randomly orientated (isotropic) sample, i.e. $\langle \cos^2 \theta \rangle = 1/3$, then eqn (13) becomes

$$\frac{1}{v_u^2} = \frac{2}{3v_\perp^2} + \frac{1}{3v_\parallel^2} \qquad (19)$$

where v_u is the ultrasonic velocity of an unorientated sample. Rearranging eqn (19) becomes

$$v_\perp^2 = 2v_u v_\parallel / 3v_\parallel^2 - v_u^2 \qquad (20)$$

Since typical values of $v_\parallel$ and v_u are of the order of 14 km/s and 1·5 km/s respectively, we can assume that $3v_\parallel^2 \gg v_u^2$ and thus

$$v_\perp^2 = 2v_u^2/3 \qquad (21)$$

Equation (21) indicates that the value of the observed velocity is directly related to the perpendicular component;[97,98] using eqn (17) we see that there will be a linear relation with the volume fraction of the crystalline phase.[99] A plot of the velocity of various polyethylene samples with linear and branched chain structures (Fig. 16), indicates that this prediction is correct and establishes the potential of the ultrasonic method as a direct non-destructive method for the determination of crystallinity in 'isotropic' samples. Thermal annealing of isotropic samples can lead to an increase in the velocity of propagation consistent with an increase in the degree of crystallinity. Investigation of the morphology reveals that this process is usually accompanied by a thickening of the lamellae from which the

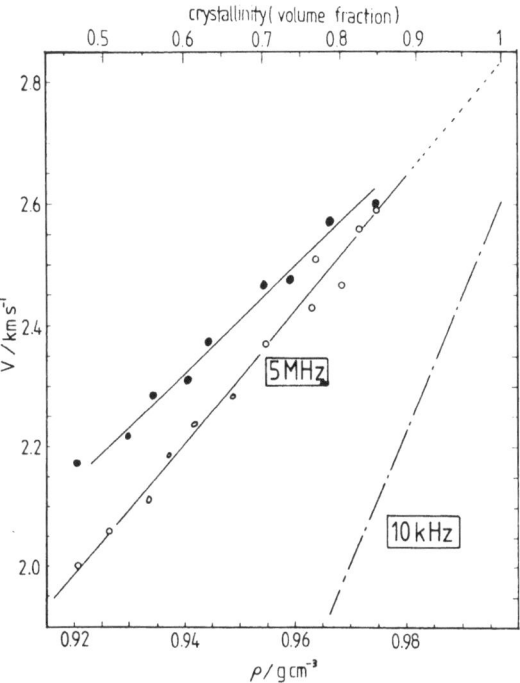

FIG. 16.   Variation of the longitudinal velocity with degree of crystallinity for linear (●) and branched chain (○) polyethylene samples.

crystallites are composed. Thermal annealing can also change the size distribution of the crystallites and this can lead to changes in the component observed when the acoustic attenuation is measured over the frequency range 1 to 1000 MHz.[99] By the use of slow cooling or quenching of polyethylene samples it is possible to generate spherulite structures with dimensions in the range 5 to 50 μm. The adiabatic passage of a sound wave through such a medium will lead to the generation of local differences in temperature (Fig. 17). Coupling between the chains forming the spherulites will be rather weak prohibiting the rapid thermal equilibration of temperature between domains which are raised to different temperatures. However, within the lamellae, which are typically 5 to 50 nm in thickness, rapid inter- and intramolecular vibrations can produce the equilibration of energy. The process by which equilibration occurs may therefore be assumed to occur via transfer from the crystallites to amorphous regions which can then interact with the surrounding crystallite structure. Calculations of the relative rates of these processes indicate that the equilibration between crystallites and the contiguous amorphous zones will occur with times in the range 10^{-7} to 10^{-9} s, whereas the gross transfer of heat between neighbouring crystallites takes approximately 10^{-4} s. The occurrence of a restriction in the rapid thermal equilibration of energy between spherulites may in part contribute to the low temperature thermal anomalies observed in the specific heat measurements. Cooling the samples will alter the

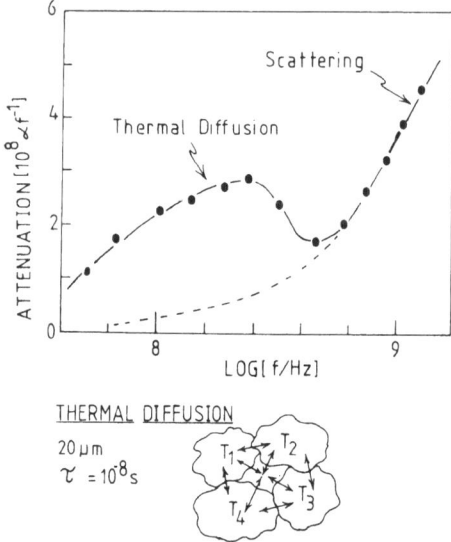

FIG. 17.   Acoustic dispersion data associated with thermal conduction between spherulites in polyethylene.

wavelength and also the number of waves (thermal phonons) available. Energy migration within a sample usually occurs via phonon–phonon interactions and hence the occurrence of a lower than statistical density of a particular type of wave can lead to a 'bottle neck' situation and hence an anomaly in the specific heat and also acoustic attenuation.

6.3. Drawn Polymers—Scattering Effects

Morphologically related phenomena other than those associated with the anisotropy of the modulus can be observed when a sample is drawn. Polypropylene has a spherulite structure which when subjected to the process of drawing will elongate and may lead to void formation. This hypothesis is supported by the observation of a minima in the density versus draw ratio[95] (Fig. 18). It is also observed that the attenuation of the drawn sample is considerably higher than that of the undrawn sample[100] (Fig. 19). Assuming that the increase in the attenuation arises principally

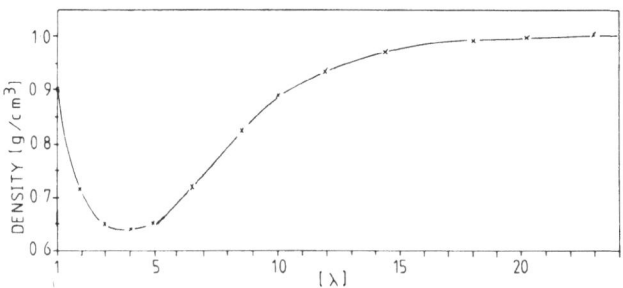

FIG. 18.   Variation of density with draw ratio for polypropylene.

from the scattering from air voids, and referring the experimental observations to the theory at room temperature, it is possible to predict the temperature dependence of the attenuation from the variation of the propagation velocity. The theory uses the fact that the mean size of the air void responsible for the scattering is approximately 0.1×1 μm which is consistent with data obtained from scanning electron microscopy (Fig. 20).

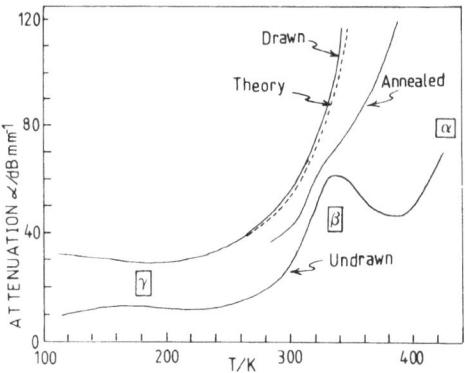

FIG. 19.   Variation of velocity and attenuation with draw ratio for polypropylene.

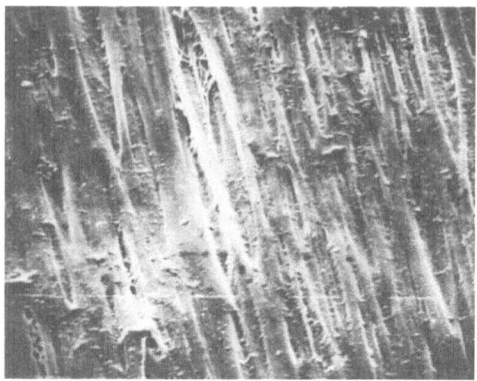

FIG. 20.   Scanning electron micrograph of drawn polypropylene. λ = [9].

7. STRESS RELAXATION

An interesting application of the ultrasonic method is in the monitoring of stress relaxation in polymer samples.[100] Investigation of the relation between the attenuation and temperature of draw indicates that for poly-propylene the lower the temperature used in drawing the higher the acoustic attenuation. This observation is consistent with the lower temperature draw material having a higher void content. Visually it is observed that samples drawn at low temperature are opaque whereas those drawn at high temperature are clear, transparent and with low acoustic attenuation. Annealing of the samples produced with high void contents leads to the observation of a decrease in the attenuation consistent with a decrease in void content. The overall length of the sample decreases on annealing and this is consistent with the reorganisation of the fibrillar structure as a consequence of internal stress. The analysis of the acoustic data indicates that the activation energy is similar to that observed for the beta process, whereas the changes in the overall length indicate a large temperature dependence. It may be suggested that the reorganisation of the void structures occurs via a relatively local motion of the polymer, whereas the motion of fibrillar structures must involve a pseudo-reptation type of motion. The ultrasonic method has considerable potential for this type of study since it does not destroy or significantly modify the sample being investigated.

8. CROSSLINKING REACTIONS

The process of cure of a variety of phenolic[101, 102] and poly(phenyl-quinoxaline)[103] materials has been investigated as a function of cure time and temperature. It was observed that in the case of the cure of phenol formaldehyde type of materials there is a significant increase in the attenuation followed by a decrease and a marked increase in the velocity as the cure proceeds. The latter can be attributed to an increase in the modulus as the network structure is formed. The systematic investigation of phenolic polymers has indicated that specimens with densities varying between 1·217 to 1·229 g/cm³ exhibited little correlation with the temperature at which cure was performed. The longitudinal velocity was observed to vary with density and not directly with the temperature at which cure was performed or degree of crosslinking achieved. As in the case of polypropylene, it appears that the acoustic properties are once more correlating with the density of voids rather than with a property of the matrix. The sound absorption, however, was found to be a function of the cure temperature, having the lowest value in the highest temperature cure; this is consistent with the lowest void content in this material. Comparison of the absorption data for phenolics and PPQ indicates that relaxation can occur in the former but not in the latter. In the case of the phenolics the magnitude of the acoustic loss was observed to vary with water content implying that the molecular relaxation process is associated with motion of the solvated ether bridge or hydrated residual phenol groups.

9. CONCLUSIONS

From this brief review of the literature it will be appreciated that the ultrasonic technique has potential both as a method for the investigation of high frequency mechanical relaxation in polymers but also for the characterisation of certain morphologically related phenomena: thermal diffusion, void scattering and stress relaxation being the most obvious.

Interest has been aroused recently in the concept of an acoustic microscope. Investigations on biological systems have indicated that this approach is viable and can yield information not available from optical systems. The high acoustic attenuation and need for resolutions in excess of 1 μm have rather limited the application of this method to polymers. It may be anticipated that improvement in definition and possibly resolution may allow this technique to be applied to the investigation of the grosser morphological features of polymers. In this context, ultrasonic measurements have been recently reported on polymer samples loaded with small glass beads. It was observed that the attenuation is directly related to the number of scatters present; however interactions between polymer and the glass substrate can in certain cases lead to shifts in the relaxation spectrum with a corresponding change in the observed attenuation.

REFERENCES

1. NORTH, A. M. and PETHRICK, R. A., in: *Development in Polymer Characterisation—2*, J. V. Dawkins, Ed., 1980, Applied Science Publishers, London, p. 183.
2. PETHRICK, R. A., *Sci. Prog.*, 1980, **66**, 571.
3. PETHRICK, R. A., *Polymer–Polymer Interactions*, Wiley Interscience, Colchester and New York (in press).
4. GILBERT, A. S., PHILLIPS, D. W. and PETHRICK, R. A., *J. Appl. Polym. Sci.*, 1977, **21**, 319.
5. NORTH, A. M., PETHRICK, R. A. and PHILLIPS, D. W., *Macromolecules*, 1977, **10**, 992.
6. WATERMAN, H. A., *Kolloid*, 1963, **192**, 1.
7. BLITZ, J., *Fundamentals of Ultrasonics, 2nd Ed.*, 1967, Plenum, New York.
8. EGGERS, F., *Acustica*, 1968, **19**, 1.
9. PHILLIPS, D. W. and PETHRICK, R. A., *J. Macromol. Sci. Rev. Macromol. Chem.* 1978, **C16**, 1.
10. HARTMAN, B., *Methods in Experimental Physics* **16C**, 1980, 59.
11. PEREPECKO, I., *Acoustic Methods of Investigating Polymers*, 1975, Mir Publications, Moscow.
12. Faraday Publications No 68, *Organization of Macromolecules in the Condensed Phase*, 1979.
13. KELLER, A. in: *Ultrahigh Modulus Polymers*, A. Cifferi and I. M. Ward, Eds, 1979, Applied Science Publishers, London, p. 321.
14. SCHMIEDER, K. and WOLF, K., *Kolloid Z.*, 1952, **127**, 65.
15. SOMMER, W., *Kolloid Z.*, 1959, **161**, 97.
16. BECKER, G. W., *Kolloid Z.*, 1955, **140**, 1.
17. PEREPECKO, I. I., TREPELKOVA, L. I., BODROVA, L. A. and BUNINA, L. A., *Vysokomol Soed.*, 1968, **10B**, 507.
18. SAUER, J. and WOODWARD, A., *Rev. Mod. Phys.*, 1960, **32**, 88.
19. JACKSON, D. A., PENTACOST, H. T. A. and POWLES, J. G., *Mol. Phys.*, 1972, **23**, 425.
20. NEKI, K. and GEIL, P. H., *J. Macromol. Sci. Phys.*, 1973, **8**, 295.
21. MASON, J. A., IOLST, S. A. and ACOSTA, R., *J. Macromol. Sci. Phys.*, 1974, **9**, 301.
22. MAEDA, Y., *J. Polym. Sci.*, 1955, **18**, 87.
23. PEZZIN, G., AJROLDI, G., CASIRAGHI, T., GORBUGHO, C. and VITTADIM, G., *J. Appl. Polym. Sci.*, 1972, **16**, 1839.

24. BUCHDAHL, R. and NIELSEN, L. E., *J. Appl. Phys.*, 1950, **21**, 482.
25. SAUER, J. A. and KLINE, D. E., *J. Polym. Sci.*, 1955, **18**, 491.
26. MAXWELL, B., *J. Polym. Sci.*, 1956, **20**, 551.
27. ILLERS, K. H. and JENCKEL, E., *Kolloid Z.*, 1959, **165**, 73.
28. YANO, O. and WADA, Y., *Rept. Progr. Polym. Phys. Japan*, 1969, **12**, 297.
29. MALKIN, A. YA., DZYURA, E. A. and VINAGRADOV, G. V., *DAN SSSR ser. khim.*, 1969, **188**, 1328.
30. POH, B. T., NORTH, A. M. and PETHRICK, R. A., *Polymer*, 1980, **21**, 772.
31. VOLKENSTEIN, M. V., *Configurational Statistics of Polymer Chains*, 1959, Mir, Moscow.
32. WOODWARD, E. A., *Pure Appl. Chem.*, 1966, **12**, 34.
33. BOYER, R. F., *Polym. Eng. Sci.*, 1968, **8**, 1.
34. ALLEN, G., HIGGINS, J. S. and WRIGHT, C. J., *Polymer*, 1972, **13**, 157.
35. NORTH, A. M., PETHRICK, R. A. and PHILLIPS, D. W., *Polymer*, 1977, **18**, 324.
36. SHIMIZU, K., YANO, O. and WADA, Y., *J. Polym. Sci. Polym. Phys.*, 1975, **13**, 1959.
37. SHIMIZU, K., YANO, O., WADA, Y. and KAWAMURA, Y., *J. Polym. Sci., Polym. Phys.*, 1973, **11**, 1641.
38. HARTMAN, B. and JARZYNSKI, J., *J. Appl. Phys.*, 1972, **43**, 4304.
39. SAITO, S. and NAKAJIMA, T., *J. Appl. Polym. Sci.*, 1959, **2**, 93.
40. THURN, H. and WOLF, W., *Kolloid Z.*, 1956, **148**, 16.
41. PEREPECKO, I. I., KVACHEVA, L. A., USHAKOV, L. A., SVETOV, A. YA, and GRECHISHKIN, V. A., *Plast. Massy*, 1970, **8**, 43.
42. WADA, Y., HIROSE, H. A., SANO, T. and FURUTOMI, S., *J. Phys. Soc. Japan*, 1959, **14**, 1064.
43. MIKHAILOV, G. P. and EIDELKANT, M. P., *Vysokomol. Soed.*, 1960, **2**, 287.
44. HARA, T., *Japan. J. Appl. Phys.*, 1967, **6**, 147.
45. HARTMAN, B. and JARZYNSKI, J., *J. Acoustic Soc. Am.*, 1974, **56**, 1469.
46. MARCINCIN, K. and ROMANOV, A., *Polymer*, 1975, **16**, 175.
47. ALLEN, G., McAINSH, J. and JEFFS, G. M., *Polymer*, 1971, **12**, 85.
48. ITO, E. and HATAKEYAMA, T., *J. Polym. Sci., Polym. Phys.*, 1975, **13**, 2313.
49. PEREPECKO, I. I. and STARTSEV, O. V., *Soviet Phys. (Acoustics)*, 1974, **20**, 456.
50. PHILLIPS, D. W., NORTH, A. M. and PETHRICK, R. A., *J. Appl. Polym. Sci.*, 1977, **21**, 1859.
51. MOWRY, S. C. and NOLLE, A. W., *J. Acoust. Soc. Am.*, 1948, **20**, 432.
52. BILLIYAR, K., BOYD, R. H., KANISKIN, V. A. and SHEN, M., *J. Polym. Sci. Polym. Phys. Ed.*, 1973, **11**, 2261.
53. NAAKE, H. J. and TAMM, K., *Acustica*, 1958, **8**, 65.
54. MASON, W. P. and McSKIMIN, H. J., *Bell Syst. Tech. J.*, 1952, 122.
55. CUNNINGHAM, J. R. and IVEY, D. G., *J. Appl. Phys.*, 1956, **27**, 967.
56. MAEDA, Y. J., *Polym. Sci.*, 1955, **18**, 87.
57. HATFIELD, P., *Br. J. Appl. Phys.*, 1950, **1**, 252.
58. BUNN, C. W. and HOWELLS, E. R., *Nature*, 1954, **174**, 549.
59. BUNN, C. W. and RIGBY, H. A., *Nature*, 1956, **164**, 583.
60. FURUKAWA, G. T., McCOSKEY, R. E. and KING, G. J., *J. Res. Natl. Bur. Stand.*, 1952, **49**, 273.
61. PEREPECKO, I. I., *Sov. Phys. Acoust.*, 1973, **18**, 343.

62. PEREPECKO, I. I. and SOROKIN, V. E., *Sov. Phys. Acoust.*, 1973, **18**, 485.
63. POH, B. T. and PETHERICK, R. A., *Mol. Phys.*, 1980, **40**, 539.
64. KOO, G. P., in: *Fluoropolymers*, L. A. Wall, Ed., 1972, Wiley, New York, p. 507.
65. KARYAKIN, N. V., RABINOVICH, I. B. and ULYANOV, V. A., *Polym. Sci. USSR*, 1969, **11**, 3159.
66. KABIN, S. P., *Sov. Phys. Acoust.*, 1956, **1**, 2542.
67. KONO, R., *J. Phys. Soc. Japan*, 1961, **16**, 1580.
68. EBY, R. K. and SINNOTT, K. M., *J. Appl. Phys.*, 1961, **332**, 1765.
69. EBY, R. K. and WILSON, F. C., *J. Appl. Phys.*, 1962, **33**, 2951.
70. SAMARA, G. A. and FRITZ, I. J., *J. Polym. Sci. Polym. Lett. Ed.*, 1975, **13**, 93.
71. MCSKIMIN, H. J., *J. Acoust. Soc. Am.*, 1951, **23**, 429.
72. LEVENE, A., PULLEN, W. J. and ROVERTS, J. M., *J. Polym. Sci.*, 1965, **A3**, 697.
73. ASSAY, J. R. and GUENTHER, A. H., *J. Appl. Polym. Sci.*, 1966, **11**, 1087.
74. WADA, Y. and YAMAMOTO, K., *J. Phys. Soc. Japan*, 1956, **11**, 887.
75. MASON, W. P. and MCSKIMIN, H. J., *Bell Syst. Tech. J.*, 1952, 122.
76. WARDEL, G., MCBRIERTY, V. J. and DOUGLAS, D. C., *J. Appl. Phys.*, 1974, **45**, 3441.
77. EBY, R. K., *J. Acoust. Soc. Am.*, 1964, **36**, 1485.
78. HARTMAN, B. and JARZYNSKI, J., *J. Acoust. Soc. Am.*, 1968, **44**, 387.
79. WATERMAN, H. A., *Kolloid Z.*, 1964, **192**, 1.
80. BORDELIUS, N. A. and SEMENCHENKO, V. K., *Sov. Phys. Acoust.*, 1971, **16**, 519.
81. HARTMAN, B. and JARZYNSKI, J., *J. Appl. Phys.*, 1972, **43**, 4304.
82. FLOCKE, H. A., *Kolloid Z.*, 1962, **180**, 118.
83. BOYER, R. F., *J. Polym. Sci. Polym. Symp.*, 1975, **50**, 189.
84. WADA, Y., HIROSE, H., ASANO, T. and FUKUTOMI, S., *J. Phys. Soc. Jap.*, 1959, **14**, 1064.
85. TAKAYANAGI, M., IMADA, K. and KAJIYAMA, T., *J. Polym. Sci.*, 1966, **15**, 263.
86. WARD, I. M. and CIFERRI, A., *Ultra High Modulus Polymers*, Applied Science Publishers, London, 1979.
87. PETHRICK, R. A. (unpublished data).
88. FLEET, J. V., MEYER, D. E., THOMAS, K. and ABRAHAMS, M., *J. Phys. D.*, 1973, **6**, 1336.
89. MOSELEY, W. W., *J. Appl. Polym. Sci.*, 1960, **3**, 266.
90. WARD, I. M., *Tex. Res. J.*, 1964, **34**, 806.
91. MORGAN, H. M., *Tex. Res. J.*, 1962, **32**, 866.
92. RIDER, J. G. and WATKINSON, K. M., *Polymer*, 1978, **18**, 845.
93. CHAN, O. K., CHEN, F. C., CHOY, C. L. and WARD, I. M., *J. Phys. D.*, 1978, **11**, 617.
94. PETHRICK, R. A. and CROFTON, D. (unpublished work).
95. DATTA, P. K. and PETHRICK, R. A., *Polymer*, 1978, **19**, 145.
96. SAMUELS, R. J., *J. Polym. Sci. A–2*, 1965, **3**, 1741.
97. URICK, R. J., *J. Appl. Phys.*, 1947, **18**, 983.
98. WATERMAN, H. A., *Kolloid Z.*, 1963, **192**, 9.
99. ADACHI, K., LAMB, J., HARRISON, G., NORTH, A. M. and PETHRICK, R. A., *Polymer*, 1981, **22**, 1026.

100. ADACHI, K., LAMB, J., HARRISON, G., NORTH, A. M. and PETHRICK, R. A., *Polymer*, 1981, **22**, 1032.
101. MARTIN, R. W., *The Chemistry of Phenolic Resins*, 1956, Wiley, New York.
102. SOTER, G. A., DIETZ, G. H. and HAUSER, E. A., *J. Ind. Eng. Chem.*, 1953, **45**, 2743.
103. HARTMAN, B., *J. Appl. Polym. Sci.*, 1975, **19**, 3241.

Chapter 6

CHARACTERISATION AND ASSESSMENT OF POLYMER ORIENTATION

HIROMICHI KAWAI

Department of Polymer Chemistry, Kyoto University, Japan

and

SHUNJI NOMURA

Department of Textile Engineering, Kyoto University of Industrial Arts and Fibre Technology, Japan

SUMMARY

The characterisation of polymer orientation is discussed rigorously in terms of the theory of Krigbaum and Roe by expanding the orientation distribution functions of certain structural units into infinite series of spherical harmonics. The orientation factors averaging the orientation distributions with respect to their moments of any given orders are also introduced in terms of the most general formulae. Also, the assessment of polymer orientation is discussed quantitatively in terms of finite series of spherical harmonics and the truncation errors. The optical anisotropy of bulk polymers is reviewed in a general way, showing how the absorption and emission dichroisms are related to the 2nd and 4th moments of polymer orientation. A graphical representation of the state of uniaxial orientation in the plots of the 2nd-order orientation factor vs the 4th-order orientation factor is proposed for assessing the uniaxial orientation behaviour of noncrystalline structural units with the 4th-order approximation.

1. INTRODUCTORY SURVEY ON POLYMER ORIENTATION STUDIES

Early works on the anisotropy in physical properties of natural fibres were focused upon form birefringence and form dichroism. Neubert[1] investigated not only intrinsic birefringence of a certain structural unit, but also the form birefringence and form dichroism in dyed and swollen fibres, and concluded that the dyes are absorbed in an oriented manner in cellulosic fibres. A similar work was carried out by Frey.[2, 3] The dichroism was already reported by Ambronn[4] for dyed cell membranes, as early as in 1888, who also suggested the absorption of dyes in the oriented manner.

Theories on the form birefringence and form dichroism in a two-phase system were developed by Wiener.[5, 6] According to his theories, a two-phase system in which rodlets orient regularly in a matrix results in form birefringence even if the rodlets and the matrix are entirely optically isotropic and absorb polarised light to an equal extent. The case of selective absorption by one of the two components in the two-phase system was called the form dichroism.[6] In the swollen cellulosic fibres investigated by Neubert, the rodlets, if any, must be intrinsically birefringent, but not isotropic. For such a swollen system of anisotropic rodlets oriented regularly in an isotropic matrix, the birefringence of the system must be formulated, as proposed by Möhring,[7] in terms of a sum of the intrinsic birefringence of the rodlets and the Wiener's form birefringence.

The first quantitative investigations on the dye-dichroism related to the micellar orientation in cellulosic fibres and films were carried out by Preston,[8] who made the first definition of dichroic ratio in the absorption of visible rays. A series of his studies[8–12] involved several kinds of optical anisotropies, such as absorption, reflection, and fluorescence of polarised light as well as birefringence, so as to assess average degree of molecular orientation within the system. Eventually, he compared the average degrees of molecular orientation derived from the different optical anisotropies with that derived from X-ray diffraction in order to evaluate crystalline and amorphous contributions to the average degrees of molecular orientation. Morey[13–15] also reported a quantitative investigation of polarised fluorescence, and defined the degree of molecular orientation in terms of the polarisation of fluorescence. However, in his method as well as in Preston's method,[11] the incident ray was not polarised in contrast to a recent fluorescence polarisation method in which polarised radiation is mostly used, as will be discussed later.

Extensive work on the optical anisotropy associated with deformation of hydrated cellulosic gel can be seen in both series of reports by Kratky[16-26] and Hermanns.[22, 27-42] Kratky studied the deformation mechanism of fibrous materials by means of X-ray diffraction and birefringence techniques, postulating two kinds of orientation models of structural elements in association with the deformation of swollen gel, and tried to explain the deformation mechanism in terms of the orientation distribution of the elements, but not of the average degree of orientation of the elements. Hermanns also studied along the same lines as those by Kratky, and introduced one of the second-order orientation factors as a measure of the average degree of orientation.[34]

Müller[43] investigated the effect of molecular orientation upon the mechanical and optical properties in fibrous as well as crosslinked rubber-like materials. Kuhn and Grün[44] proposed a model of the so-called freely joined equivalent chain which has been highly successful in furnishing quantitative descriptions of physical properties of crosslinked rubber-like materials under deformation, such as entropic retractive force and birefringence, as discussed by Treloar.[45] Mechanical anisotropy in pre-oriented systems was discussed by Raumann[46-48] and Ward et al.[49-53] in terms of the complete tensor of elastic compliance on the basis of the infinitesimal deformation theory of an anisotropic elastic body. Ward formulated the mechanical and optical anisotropies of uniaxially oriented materials in terms of the degree of orientation of a certain kind of structural unit and its intrinsic anisotropies on the basis of the aggregation model,[49, 54] and further extended his studies to the viscoelastic anisotropy.[54-57] Similar kinds of studies on the mechanical and optical anisotropies as well as swelling anisotropy were performed by Nomura et al.[58, 59] and Maeda et al.[60] for two-phase systems of semicrystalline polymers.

Recently, the measurements of molecular orientation have been much developed technically and theoretically. Quantitative discussion on visible absorption dichroism can be extended to ultraviolet (UV) and infrared (IR) dichroisms without any change of principle, but only with the change of structural unit absorbing the radiated ray, as fully discussed by Nomura et al.[61] That is, at a given wave number of IR radiation, the structural unit must be a specific chemical bond or group. Quantitative description of IR dichroism was given by Fraser[62] and Beer[63] in terms of the molecular orientation and the direction of the transition moment within the molecule. IR dichroism in biaxially oriented systems was discussed by Stein[64] and Koenig.[65] UV and visible dichroisms for conjugated double-bonds were also discussed by Okajima et al.[66] and Shindo and Stein.[67, 68] The absorption dichroisms, all mentioned above, are parameters which are able to determine the second moments of the orientation distribution of the corresponding structural units, i.e. the second-order orientation factors averaging the orientation of the units.

More recently, the measuring technique of emission dichroisms, such as fluorescence polarisation and Raman polarisation, as well as nuclear magnetic resonance (NMR) have also been much developed. The fluorescence polarisation technique was pioneered by Nishijima and his coworkers[69–71] and was discussed by several authors.[72–77] The Raman polarisation can be discussed in terms of a common concept with the emission dichroism of the fluorescence polarisation.[75, 78, 79] Differences between them are what kind of electromagnetic wave is employed and what kind of structural unit is concerned with the absorption and emission of the wave. All of these emission dichroisms can give the second and fourth moments of the orientation distribution of the corresponding structural units, i.e. the second- and fourth-order orientation factors. Further, the NMR technique can give in principle the higher-order moments up to the eighth order,[74, 80–85] but with considerable difficulties in practice. Table 1 lists various kinds of measuring techniques and the information which is available for the determination of the orientation factors. Detailed discussions on the orientation measurements were covered in several reviews.[86–88]

As recognised from the table, the optical methods of absorption and emission dichroisms give the second and fourth moments of the orientation distribution of the corresponding structural units, while X-ray diffraction gives the orientation distribution of the reciprocal lattice vectors of a given crystal plane and, subsequently, every order of the moment. Strictly speaking, however, all of these experimental methods give only information about the orientation of a particular axis fixed within the structural unit, but not of the unit itself which must be a volume element of three dimensions. Therefore, a phenomenological theory proposed by Krigbaum and Roe[89, 90] and further extended by Nomura et al.,[91] is extremely significant for systematising the characterisation and assessment of polymer orientation. That is, Krigbaum and Roe defined two types of orientation distribution functions for the structural unit of three dimensions and a particular axis fixed within the structural unit, respectively, and expanded the distribution functions into infinite series of spherical harmonics of three and two indices in order to interrelate the coefficients of the expanded series, and the moments of the orientation distributions, with each other by the aid of the Legendre addition theorem and, further, the distribution functions themselves.

TABLE 1

VARIOUS METHODS OF ORIENTATION MEASUREMENT AND THE
INFORMATION THEY CAN PROVIDE

Method	Structural element related to the method	Orientation factors	Available references
Birefringence	Contributions cannot be separated into concrete structural elements without any other information.	2nd order	86–88, 96
Visible dyed dichroism	Non-crystalline region: it is assumed that dyes are absorbed parallel or with a given geometry to a non-crystalline chain segment.	2nd order	12, 97
Visible or UV dichroism	Conjugated double bonds or polymer segments in non-crystalline region.	2nd order	67, 68
IR dichroism	Molecular bond or simple group in both crystalline and non-crystalline regions, separately.	2nd order	65, 98
Laser–Raman spectroscopy	Same structural elements as those in IR dichroism	2nd and 4th orders	75, 78, 79
Fluorescence polarisation	Non-crystalline region; similar assumption as that for dyed dichroism.	2nd and 4th orders	69–77
Broad-line NMR	Internuclear vector: possible to distinguish crystalline and non-crystalline contributions.	up to 4th order $\langle \Delta H^2 \rangle$ up to 8th order $\langle \Delta H^4 \rangle$	80–85
Wide-angle X-ray diffraction	Reciprocal lattice vector of crystal.	Complete orientation distribution function	89, 90, 99, 100

Graphical representations of the state of orientation of the structural units in terms of a limited number of orientation factors or moments have been proposed by several authors with corresponding orders of approximation. Stein *et al.*[92] proposed orthogonal–equilateral triangle coordinates representing the state of uniaxial orientation of orthorhombic crystals in terms of the second-order orientation factors of the principal crystallographic axes. Desper and Stein[93, 94] proposed equilateral triangle coordinates representing the state of biaxial orientation of a particular crystallographic axis in terms of its second moments of orientation with respect to cartesian coordinate axes fixed within the specimen space, and Nomura et al.[95] proposed orthogonal coordinates representing the state of uniaxial orientation of a particular axis fixed within the structural unit in terms of the second- and fourth-order orientation factors.

In this chapter, let us first introduce the theory of Krigbaum and Roe in order to generalise the description of orientation of structural units and the orientation factor as well. In turn, let us assess the orientation distribution function of crystalline structural units of three dimensions from measurable quantities, i.e. finite series of spherical harmonics, and discuss the truncated error quantitatively. Second, the absorption and emission behaviour of optically anisotropic units will be discussed in order to unify the absorption and emission dichroisms as well as birefringence for determining the second- and fourth-order moments of orientation. A graphical representation of the state of uniaxial orientation proposed by Nomura et al.[95] in terms of the second- and fourth-order moments will be introduced for assessing the uniaxial orientation of noncrystalline chain segments with the fourth-order approximation. Finally, a study on the deformation mechanism of spherulitic crystalline polymers will be demonstrated as one of the applications of the theory of Krigbaum and Roe for interrelating two types of orientation distribution functions, i.e. the orientation distribution function of the reciprocal lattice vector of a given crystal plane as a function of lamellar orientation associated with deformation of spherulitic crystalline texture.

2. ORIENTATION DISTRIBUTION FUNCTION AND ORIENTATION FACTOR

Consider Fig. 1 which shows a structural element in a fixed laboratory cartesian coordinate system $0-x_1x_2x_3$. In general this fixed frame of reference is associated with the principal extension ratio directions of the sample, e.g. x_3 is parallel to the principal extension axis (machine or draw axis) while x_2 and x_1 are along the transverse and film normal directions, respectively. Now let us place within this frame a second cartesian coordinate system $0-u_1u_2u_3$. As an example case, the structural element might be an orthorhombic crystal such that one might then choose to let u_1, u_2 and u_3 lie parallel to the a, b and c axes of this crystal. However, for the purpose of describing any type of element, let us only state that the $0-u_1u_2u_3$ frame is fixed within the element—the choice of the spatial arrangement of $0-u_1u_2u_3$ will depend on the nature of the geometry or symmetry of the element but its selection need not be unique except for the u_3 axis being possibly parallel to molecular axis of the element.

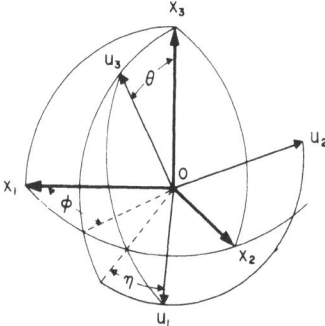

FIG. 1. Euler angles, ϕ, θ, and η, specifying the orientation of cartesian co-ordinates $0-u_1u_2u_3$ fixed within a structural element with respect to other cartesian coordinates $0-x_1x_2x_3$ fixed within the space of a bulk sample. x_1, thickness direction; x_2, transverse direction; x_3, machine direction.

From Fig. 1 we now ask how can one specify the orientation of the element (i.e. the orientation of $0-u_1u_2u_3$) relative to the fixed laboratory frame $0-x_1x_2x_3$? Clearly specification of the three Euler angles, ϕ, θ, and η, shown will permit this description. That is, θ and ϕ, the polar and azimuthal angles, denote the orientation of u_3 with respect to $0-x_1x_2x_3$. The orientation of the element is now fixed once the rotation angle η about the u_3 axis is given. If many of these same elements existed within a sample space under investigation, one might desire to determine the number in any selected spatial position specified by the angular range $\theta-\theta+d\theta$, $\phi-\phi+d\phi$, and $\eta-\eta+d\eta$. Clearly, this number at any angular position could be represented by an orientation distribution function $w(\theta,\phi,\eta)$ such that $w(\theta,\phi,\eta)\sin\theta\,d\theta\,d\phi\,d\eta$ will represent the probability of finding a structural element in this angular range. For the above to hold, the distribution function may be reformed for convenience and normalised as follows:

$$\int_0^{2\pi} \int_0^{2\pi} \int_0^{\pi} w(\theta, \phi, \eta) \sin \theta \; d\theta \; d\phi \; d\eta =$$

$$\int_0^{2\pi} \int_0^{2\pi} \int_{-1}^{1} w(\xi, \phi, \eta) \; d\xi \; d\phi \; d\eta = 1 \quad (1)$$

where ξ is defined as equal to $\cos \theta$.

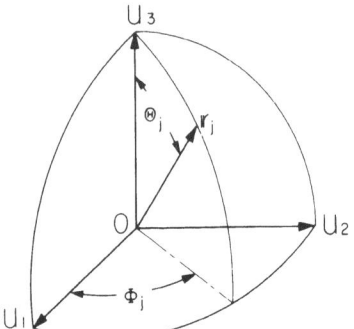

FIG. 2. Angles Θ_j and Φ_j specifying the orientation of a given jth vector of the structural element with respect to the cartesian coordinates $0-u_1 u_2 u_3$ fixed within the element.

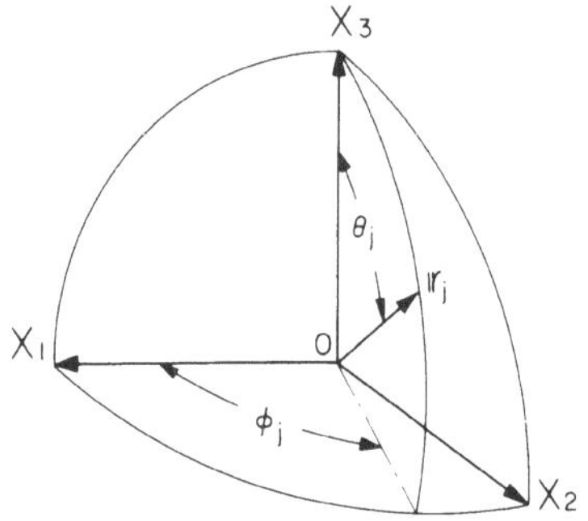

FIG. 3. Angles θ_j and ϕ_j specifying the orientation of a given jth vector of the structural element with respect to the cartesian coordinates $0-x_1 x_2 x_3$ fixed within the space of the bulk sample.

While the above has led to a description of the orientation of $0-u_1 u_2 u_3$ (the *volume* element) with respect to the laboratory frame $0-x_1 x_2 x_3$, any experiment never allows a direct measurement of the orientation of the volume element in terms of the rigorous distribution of the three Euler angles. Rather it may only allow the determination of the orientation of an axis along another direction within the element. For example, Fig. 2 shows a vector $\mathbf{r}_j$ which is *within the element* and whose orientation within this element (i.e. relative to $0-u_1 u_2 u_3$) is specified by polar and azimuthal angles, Θ_j and Φ_j, whereas its orientation relative to the laboratory frame $0-x_1 x_2 x_3$ is given by the other polar and azimuthal angles, θ_j and ϕ_j, as shown in Fig. 3. If indeed it is the orientation distribution of $\mathbf{r}_j$ (or its average) that is determined by experiment, then one must know the transformation relationship between the two different orientation distributions and their averages quantitatively if the orientation of the structural element is to be specified.

The two sets of angles, (θ_j, ϕ_j) and (Θ_j, Φ_j) referring to the identical vector r_j, can be related to each other through a linear transformation of the two cartesian coordinate systems accompanied by the rotation of the coordinate axes by ϕ, θ, and η, and the interrelation between the seven angles defined can be given by

$$
\begin{pmatrix} \sin \theta_j \cos \phi_j \\ \sin \theta_j \sin \phi_j \\ \cos \theta_j \end{pmatrix} = (t_{ki}) \begin{pmatrix} \sin \Theta_j \cos \Phi_j \\ \sin \Theta_j \cos \Phi_j \\ \cos \Theta_j \end{pmatrix} \tag{2}
$$

where (t_{ki}) is a linear transformation operator (the direction cosine of each of the u_k axes with respect to each of the x_i axes).

Generally, the experimental measurements provide either a measure of the spatial orientation distribution of various r_j vectors (e.g. the reciprocal lattice vector of the jth crystal plane) as obtained by wide angle X-ray diffraction or only the second and/or fourth moments of the orientation distribution function for a given vector r_j, $q_j(\theta_j, \phi_j)$, as obtained from measurements of absorption and/or emission dichroisms. The distribution function $q_j(\theta_j, \phi_j)$ can be defined, similarly to eqn (1), as follows:

$$
\int_0^{2\pi} \int_0^{\pi} q_j(\theta_j, \phi_j) \sin \theta_j \, d\theta_j \, d\phi_j =
$$

$$
\int_0^{2\pi} \int_{-1}^{1} q_j(\zeta_j, \phi_j) \, d\zeta_j d\phi_j = 1 \tag{3}
$$

where ζ_j is defined as equal to $\cos \theta_j$.

One of the most fundamental problems for describing, characterising, and assessing the orientation of the structural element, is how to relate the orientation distribution of the element, $w(\xi, \phi, \eta)$, to the measurable quantities, $q_j(\zeta_j, \phi_j)$ or its moments up to, at least, the fourth moments. This can be done quantitatively, as fully discussed by Roe and Krigbaum[89, 90] and by Nomura and Kawai,[91, 95] by a well established mathematical procedure expanding both of the distribution functions into infinite series of spherical harmonics and interrelating the coefficients of the expanded series to each other with the aid of Legendre's addition theorem. Let us show the procedure as briefly as possible and introduce some important results.

Taking into account the orthogonal properties of the spherical harmonics and the normalisation conditions given by eqns (1) and (3), the orientation distribution functions can be expanded into infinite series of spherical harmonics as follows:

$$w(\xi, \phi, \eta) = \sum_{l=0}^{\infty} \sum_{m=-l}^{l} \sum_{n=-l}^{l} [A_{lmn} \cos(m\phi + n\eta)$$

$$+ B_{lmn} \sin(m\phi + n\eta)] Z_{lmn}(\xi) \qquad (4)$$

$$q_j(\zeta_j, \phi_j) = \sum_{l=0}^{\infty} \sum_{m=-l}^{l} [\alpha_{lm}^j \cos(m\phi_j) + \beta_{lm}^j \sin(m\phi_j)] \Pi_l^m(\zeta_j) \qquad (5)$$

where $Z_{lmn}(x)$ and $\Pi_l^m(x)$ are the normalised associated Legendre's polynomials having three and two variables, respectively, A_{lmn}, B_{lmn}, α_{lm}^j, and β_{lm}^j are the coefficients of the expanded series, and l, m, and n are indices (integers) showing the order of the expansion.

The function $Z_{lmn}(x)$ is given by

$$Z_{mn}(x) = \left[\frac{2l+1}{2} \frac{(l-m)!}{(l-n)!} \frac{(l+m)!}{(l+n)!} \right]^{1/2} (1/2)^m$$

$$\times (1-x)^{(m-n)/2} (1+x)^{(m+n)/2} P_{l-m}^{(m-n,m+n)}(x) \qquad (6)$$

where $P_{l-m}^{(m-n,m+n)}(x)$ is Jacobi's polynomial. Similarly, the function $\Pi_l^m(x)$ is given by

$$\Pi_l^m(x) = Z_{lm0}(x) = \left[\frac{2l+1}{2} \frac{(l-m)!}{(l+m)!} \right]^{1/2} P_l^m(x) \qquad (7)$$

where $P_l^m(x)$ is an associated Legendre's polynomial (not normalised).

In turn, the coefficients of the expanded series can be represented by

$$\left. \begin{matrix} A_{lmn} \\ B_{lmn} \end{matrix} \right\} = \int_0^{2\pi} \int_0^{2\pi} \int_{-1}^{1} w(\xi, \phi, \eta) Z_{lmn}(\xi)$$

$$\left. \begin{matrix} \cos(m\phi + n\eta) \\ \sin(m\phi + n\eta) \end{matrix} \right\} \, d\xi \, d\phi \, d\eta / 4\pi^2 \qquad (8)$$

$$\left. \begin{matrix} \alpha_{lm}^j \\ \beta_{lm}^j \end{matrix} \right\} = \int_0^{2\pi} \int_{-1}^{1} q_j(\zeta_j, \phi_j) \Pi_l^m(\zeta_j) \left. \begin{matrix} \cos(m\phi_j) \\ \sin(m\phi_j) \end{matrix} \right\} \, d\zeta_j \, d\phi_j / 2\pi \qquad (9)$$

Clearly from the above equations, the results to be emphasised are the fact that the coefficients, (A_{lmn} and B_{lmn}) and (α_{lm} and β_{lm}), are the averages of the distribution functions with respect to the lmnth and lmth orders of the spherical harmonics, respectively. In other words, the coefficients are explicitly the lmnth and lmth orders of the moments of the orientation distribution functions, respectively, and are very significant for defining the orientation factors, in general, as will be discussed later. Applying a generalisation of the Legendre addition theorem to eqn (2), the following relation may be obtained;

$$\Pi_l^m(\zeta_j) \exp(im\phi_j) = \left[\frac{2}{2l+1}\right]^{1/2} \sum_{n=-l}^{l} Z_{lmn}(\xi) \exp[i(m\phi + n\eta)]$$

$$\times \Pi_l^m(\cos\Theta_j) \exp(in\Phi_j) \quad (10)$$

Multiplying both sides of eqn (10) by $w(\xi, \phi, \eta) \cdot q_j(\zeta_j, \phi_j)$, integrating over the ranges of the five angles, and taking the relations of eqns (8) and (9) into account, then the following important relation may be obtained:

$$\left.\begin{array}{c}\alpha_{lm}^j \\ \beta_{lm}^j\end{array}\right\} = 2\pi \left(\frac{2}{2l+1}\right)^{1/2} \sum_{n=-l}^{l}$$

$$\left\{\begin{array}{c}A_{lmn}\cos(n\Phi_j) - B_{lmn}\sin(n\Phi_j) \\ A_{lmn}\sin(n\Phi_j) + B_{lmn}\cos(n\Phi_j)\end{array}\right\} \quad \Pi_l^m(\cos\Theta_j) \quad (11)$$

Equation (11) relates not only the coefficients, (A_{lmn} and B_{lmn}) and (α_{lm}^j and β_{lm}^j), to each other but also the distribution functions, $w(\xi, \phi, \eta)$ and $q_j(\zeta_j, \phi_j)$, to each other, quantitatively. Thus, the coefficients, A_{lmn} and B_{lmn}, as well as the distribution function, $w(\xi, \phi, \eta)$, may be evaluated from $q_j(\zeta, \phi_j)$ or its moments, α_{lm}^j and β_{lm}^j, which are measurable quantities as discussed above.

The functions, $Z_{lmn}(x)$ and $\Pi_l^m(x)$, have symmetries with respect to m, n and x. The orientation distribution function, $q_j(\zeta_j, \phi_j)$ also has symmetry, such as orthogonal–biaxial symmetry as given by $q_j(\pm\zeta_j, \pm\phi_j) = q_j(\pm\zeta_j, \pi \pm \phi_j)$ or uniaxial symmetry as given by $q_j(\zeta_j, \phi_j) = q_j(\zeta_j, 0)$ for most of the industrial products of sheet-like or fibrous materials, respectively, providing that the laboratory coordinate system $0–x_1x_2x_3$ is properly fixed within the sample space, as suggested in Fig. 1. In addition, the structural element itself has symmetry, such as orthorhombic or cylindrical symmetry with respect to the u_3 axis. These symmetries make the general description given by eqns (4), (5) and (11) much simpler, mainly because the coefficients having an odd number of any l, m or n vanish from the equations and partly because of the further symmetrical nature of the remaining coefficients with respect to the sign of m and n. The reader who is interested in more details may wish to refer to the references 89–91, 95, especially the paper by Roe.[90]

Let us show some simplified results under combinations of particular symmetries of the distribution function q_j and the structural element. Orthogonal–biaxial symmetry of q_j makes the coefficients *not zero* only when l and m are *even numbers*. Uniaxial symmetry of q_j also makes the coefficients *not zero* only when l is an *even number* and m is *zero*. Orthorhombic symmetry of the element makes the coefficients *not zero* only when n is an *even number*. Cylindrical symmetry of the element makes the coefficients *not zero* only when n is *zero*. As a result, eqns (4) (5) and (11) are simplified as follows:

For the orthogonal–biaxial symmetry of q_j with the orthorhombic element;

$$w(\xi,\phi,\eta) = \sum_{l=0}^{\infty} A_{l00}\Pi_l^0(\xi)$$

$$+ 2\sum_{l=2}^{\infty}\sum_{m=2}^{l} A_{lm0}\Pi_l^m(\xi)\cos(m\phi) + 2\sum_{l=2}^{\infty}\sum_{n=2}^{l} A_{l0n}\Pi_l^n(\xi)\cos(n\eta)$$

$$+ 2\sum_{l=2}^{\infty}\sum_{m=2}^{l}\sum_{n=2}^{l} A_{lmn}[Z_{lmn}(\xi)\cos(m\phi+n\eta) + Z_{lm\bar{n}}(\xi)\cos(m\phi-n\eta)] \quad (12)$$

$$q_j(\zeta_j,\phi_j) = \sum_{l=0}^{\infty}\alpha_{l0}^j\Pi_l^0(\zeta_j) + 2\sum_{l=2}^{\infty}\sum_{m=2}^{l}\alpha_{lm}^j\Pi_l^m(\zeta_j)\cos(m\phi_j) \quad (13)$$

$$\alpha_{lm}^j = 2\pi\left(\frac{2}{2l+1}\right)^{1/2}[A_{lm0}\Pi_l^0(\cos\Theta_j)$$

$$+ 2\sum_{n=2}^{l} A_{lmn}\Pi_l^n(\cos\Theta_j)\cos(n\Phi_j) \quad (14)$$

For the orthogonal–biaxial symmetry of q_j with the cylindrical element;

$$w(\xi,\phi,\eta) = \sum_{l=0}^{\infty} A_{l00}\Pi_l^0(\xi) + 2\sum_{l=2}^{\infty}\sum_{m=2}^{l} A_{lm0}\Pi_l^m(\xi)\cos(m\phi) \quad (15)$$

$$q_j(\zeta_j,\phi_j) = \sum_{l=0}^{\infty}\alpha_{l0}^j\Pi_l^0(\zeta_j) + 2\sum_{l=2}^{\infty}\sum_{m=2}^{l}\alpha_{lm}^j\Pi_l^m(\zeta_j)\cos(m\phi_j) \quad (16)$$

$$\alpha_{lm}^j = 2\pi\left(\frac{2}{2l+1}\right)^{1/2} A_{lm0}\Pi_l^0(\cos\Theta_j) \quad (17)$$

For the uniaxial symmetry of q_j with the orthorhombic element;

$$w(\xi,\phi,\eta) = \sum_{l=0}^{\infty} A_{l00}\Pi_l^0(\xi) + 2\sum_{l=2}^{\infty}\sum_{n=2}^{l} A_{l0n}\Pi_l^n(\xi)\cos(n\eta) \quad (18)$$

$$q_j(\zeta_j,\phi_j) = \sum_{l=0}^{\infty}\alpha_{l0}^j\Pi_l^0(\zeta_j) \quad (19)$$

$$\alpha_{l0}^j = 2 \left(\frac{2}{2l+1}\right)^{1/2} [A_{l00}\Pi_l^0 (\cos\Theta_j) + 2\sum_{n=2}^{l} A_{l0r}\Pi_l^n(\cos\Theta_j)\cos(n\Phi_j) \quad (20)$$

As has been recognised from eqns (8) and (9), the expansion coefficients, (A_{lmn} and B_{lmn}) and (α_{lm}^j and β_{lm}^j) may be understood as averages of the orientation distribution functions, $w(\xi, \Phi, \eta)$ and $q_j(\zeta_j, \phi_j)$, with respect to the lmnth and lmth orders of the spherical harmonics, i.e. the lmnth and lmth order moments of the distribution functions, respectively, which have been called 'orientation functions' or 'orientation factors'. Therefore, the orientation factors can be defined, in general, in terms of the coefficients as follows:[91, 95]

$$f_{lmn} = \left[\frac{2}{2l+1}\frac{(l+m)!}{(l-m)!}\frac{(l+n)!}{(l-n)!}\right]^{1/2} 4\pi^2(A_{lmn} + iB_{lmn}) \quad (21)$$

$$f_{lm}^j = \left[\frac{2}{2l+1}\frac{(l+m)!}{(l-m)!}\right]^{1/2} 2\pi(\alpha_{lm}^j + i\beta_{lm}^j) \quad (22)$$

where the coefficients in the imaginary component, B_{lmn} and β_{lm}^j, vanish away to *zero* under the symmetrical conditions of the distribution function and the structural element, as discussed above.

f_{l00} and f_{l0}^j are the $l00$th and $l0$th order orientation factors specifying the average orientation of the u_3 axis and the $\mathbf{r}_j$ vector, respectively, both with respect to the reference axis x_3. The distribution of the angles, θ and θ_j, only are taken into consideration, while those of the other angles, ϕ and η and ϕ_j, are just smeared out. Therefore, the factors must be unique only when the distribution functions have uniaxial symmetry with respect to the x_3 axis and when for f_{l00} the structural element has cylindrical symmetry with respect to the u_3 axis. f_{lm0} and f_{lm}^j are the $lm0$th and lmth order orientation factor specifying the average biaxial orientation of the u_3 axis and the $\mathbf{r}_j$ vector within the $0-x_1x_2x_3$ coordinate system. Further, f_{l0n} is the $l0n$th order orientation factor specifying the average orientation of the element $0-u_1u_2u_3$ with respect to the reference axis x_3, i.e. in turn, an average biaxial orientation of the x_3 axis with respect to the $0-u_1u_2u_3$ coordinates.

Actually, some of the generalised orientation factors defined by eqns (21) and (22) have been proposed by several authors, independently. The *second order* orientation factors f_{200} and f_{20}^j have been proposed by Hermanns and Platzek[34] and by Stein and Norris,[99, 101] respectively, for relating the birefringence of fibrous materials to the uniaxial orientation of transversely isotropic (cylindrical-symmetric) structural elements and for characterising the uniaxial orientation of polyethylene crystals, with an orthorhombic element, in terms of the average orientation of its principal crystallographic axes, a, b, and c axes. f_{220} and f'_{22} are also the *second order* orientation factors defined by Stein[96] and Nomura *et al.*[61] for relating the absorption dichroism of a material to the biaxial orientation of the structural elements. f_{400} and f'_{40} and the *fourth order* orientation factors defined by Kimura *et al.*[73] for relating the emission dichroism of a material to the uniaxial orientation of the structural elements.

Some of the generalised orientation factors can be derived from eqns (21) and (22) in more concrete forms as follows:

$$f_{200} = \sqrt{2/5}\, 4\pi^2 A_{200} = (1/2)\,(3\langle \cos^2\theta \rangle - 1) \tag{23}$$

$$f_{220} = (12/\sqrt{15})\, 4\pi^2 A_{220} = 3\langle \sin^2\theta \cos 2\phi \rangle \tag{24}$$

$$f_{202} = (12/\sqrt{15})\, 4\pi^2 A_{202} = 3\langle \sin^2\theta \cos 2\eta \rangle \tag{25}$$

$$f_{400} = (\sqrt{2}/3)\, 4\pi^2 A_{400} = (35\langle \cos^4\theta \rangle - 30\langle \cos^2\theta \rangle + 3)/8 \tag{26}$$

$$f_{420} = (4\sqrt{5})\, 4\pi^2 A_{420} = (15/2)\, \langle (7\cos^2\theta - 1)\sin^2\theta \cos 2\phi \rangle \tag{27}$$

$$f_{440} = (16\sqrt{35})\, 4\pi^2 A_{440} = 105\langle \sin^4\theta \cos 4\phi \rangle \tag{28}$$

$$f'_{20} = \sqrt{2/5}\, 2\pi\, \alpha'_{20} = (3\langle \cos^2\theta_j \rangle - 1)/2 \tag{29}$$

$$f'_{22} = (12/\sqrt{15})\, 2\pi\, \alpha'_{22} = 3\langle \sin^2\theta_j \cos 2\phi_j \rangle \tag{30}$$

$$f'_{40} = (\sqrt{2}/3)\, 2\pi\, \alpha'_{40} = (35\langle \cos^4\theta_j \rangle - 30\langle \cos^2\theta_j \rangle + 3)/8 \tag{31}$$

$$f'_{42} = (4\sqrt{5})\, 2\pi\, \alpha'_{42} = (15/2)\langle (7\cos^2\theta_j - 1)\sin^2\theta_j \cos 2\phi_j \rangle \tag{32}$$

$$f'_{44} = (16\sqrt{35})\, 2\pi\, \alpha'_{44} = 105\langle \sin^4\theta_j \cos 4\phi_j \rangle \tag{33}$$

where the angle brackets $\langle \; \rangle$ mean average.

By deriving the relationship between (A_{lmn} and B_{lmn}) and (α'_{lm} and β'_{lm}), as given by eqn (14), under a less limited condition of the orthogonal–biaxial symmetry of q_j alone being not associated with further symmetry of the structural element, like either orthorhombic or cylindrical symmetry, it is possible to obtain a more general relation than that given by eqn (14):

$$\alpha'_{lm} = 2\pi \left(\frac{2}{2l+1} \right)^{1/2} [A_{lm0}\Pi_l^0(\cos\Theta_j)$$

$$+ 2\sum_{n=1}^{l} \{A_{lmn}\cos(n\Phi_j) - B_{lmn}\sin(n\Phi_j)\}\, \Pi_l^n(\cos\Theta_j)] \tag{34}$$

When eqns (21) and (22) are substituted into eqn (34), eqn (34) can be rewritten in terms of the generalised orientation factors, f_{lm}^j and f_{lmn}, as follows:

$$f_{lm}^j = f_{lm0}P_l(\cos \Theta_j) + 2 \sum_{n=1}^{l} \frac{(l-n)!}{(l+n)!} P_l^n(\cos \Theta_j)$$
$$\times \text{Re}[f_{lmn} \exp(in\Phi_j)] \quad (35)$$

where l and m are even numbers, and $\text{Re}[X]$ is the real component of the complex quantity of X. Equation (35) must be useful for determining the degree of orientation of a particular reciprocal lattice vector, $f_{lm}^{j_0}$ from those of the other vectors $(j \neq j_0)$, as suggested by Wilchinski,[102] Sack,[103] and Krigbaum and Roe.[89, 90, 104] In turn, in order to determine f_{lmn} from f_{lm}^j for given l and m, it is necessary to measure f_{lm}^j under varying j from one to $(2l + 1)$ and to solve the simultaneous equations, as deduced from eqn (35) with respect to f_{lmn}.

For the system of orthogonal–biaxial symmetry with orthorhombic elements, the following relation simpler than eqn (35) can be obtained from eqn (14):

$$f_{lm}^j = f_{lm0}P_l(\cos \Theta_j) + 2 \sum_{n=2}^{l} \frac{(l-n)!}{(l+n)!} f_{lmn}P_l^n(\cos \Theta_j) \cos(n\Phi_j) \quad (36)$$

where l, m, and n must be even numbers, and the number of f_{lm}^j to be measured for determining f_{lmn} may be reduced to $(l/2 + 1)$. For the system of uniaxial symmetry with orthorhombic elements, the following similar relation to eqn (36) can be obtained from eqn (20):

$$f_{l0}^j = f_{l00}P_l(\cos \Theta_j) + 2 \sum_{n=2}^{l} \frac{(l-n)!}{(l+n)!} f_{l0n}P_l^n(\cos \Theta_j) \cos(n\Phi_j) \quad (37)$$

Figure 4 shows contour plots of $w(\cos \theta, \phi, \eta)$ of polyethylene crystals (orthorhombic elements) for a biaxially stretched film with extension ratios of $1 \cdot 7 \times 1 \cdot 6$, calculated by Krigbaum, Adachi and Dawkins[105] using eqn (36) with $l \leq 12$ and with the number of $q_j(\zeta_j, \phi_j)$, i.e. the number of crystal planes to be measured, being up to 13. The laboratory coordinate system $0 - x_1 x_2 x_3$ is fixed as x_3 and x_2 being parallel to the higher and lower extension ratio directions, and the $0 - u_1 u_2 u_3$ coordinates of the structural element is fixed as u_1, u_2, and u_3 being parallel to the a, b, and c axes of the orthorhombic crystal. Therefore, the most characteristic behaviour revealed from the figure are the ridge of high probability along $\phi = 90°$ and $\eta = 0°$ and almost parallel contour lines to the ridge at $\phi = 90°$. These

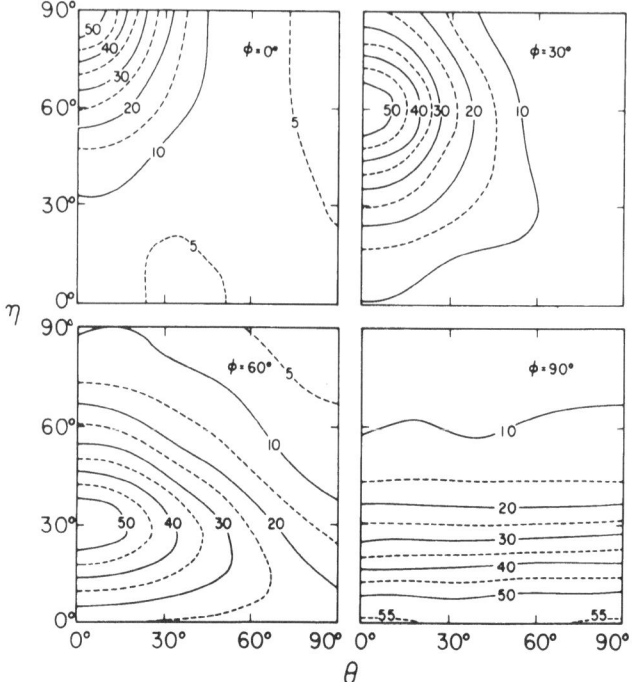

FIG. 4. Contour plots of $w(\cos\theta, \phi, \eta)$ of polyethylene crystals (orthorhombic elements) for a biaxially stretched film with extension ratios of 1·7 along the x_3 axis and 1·6 along the x_2 axis. (From Krigbaum *et al.*)[105]

indicate that the b axis has a preferential orientation along the normal to the film surface, while the a and c axes lie in the plane parallel to the film surface and are more or less randomly arranged in that plane to give nearly cylindrical symmetrical properties of the orientation distribution around the film normal. The orientation behaviour may be explained, as suggested by Takahara *et al.*[106] for biaxially stretched polypropylene, in terms of a biaxial deformation mechanism of spherulitic crystalline texture, i.e. lamellar orientation toward the film surface in association with reorientation of crystal grains within the orienting and disintegrating lamellae.

Figure 5 also shows contour plots of $w(\cos\theta, 0, \eta)$ of polyethylene crystals for a uniaxially stretched film with extension ratios up to 1·4, calculated by Fujita *et al.*[107] using eqn (37) with $l \leq 24$ and with j up to 13 for obtaining f'_{l0} from measured quantities, $q_j(\zeta_j, 0)$. The function $w(\cos\theta, 0, \eta)$ shows two peaks at $\theta = 20°$ and $\eta = 0°$ and at $\theta = 35°$ and $\eta = 90°$. This suggests two populations of crystal orientations, which may

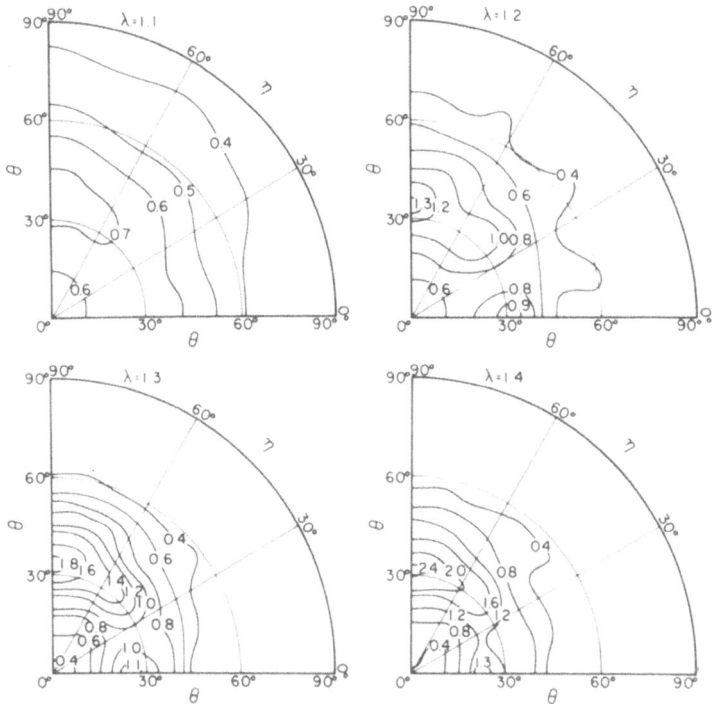

FIG. 5.   Contour plots of $w(\cos\theta, 0, \eta)$ of polyethylene crystals for a uniaxially stretched film with extension ratios up to 1·4 along the x_3 axis. (From Fujita *et al.*)[107]

be explained again in terms of a spherulite deformation mechanism, i.e. uniaxial orientation of crystal lamellae in association with reorientation of crystal grains within the orienting and disintegrating lamellae, as proposed by several authors.[108–112] The former peaks at $\theta = 20°$ and $\eta = 0°$ may arise from the crystal reorientation mostly at the equatorial zone of the uni-axially deformed spherulites, such that the reorientation is represented by a preferential rotation of the crystal grain around its own b axis (lamellar axis) so as to incline the c axis (molecular axis) parallel to the plane including the stretching direction and the lamellar axis, i.e. lamellar untwisting due to straining of the so-called tie-chain molecules between adjacent lamellae. On the other hand, the latter peak at $\theta = 35°$ and $\eta = 90°$ may arise from the crystal reorientation at polar and/or 45° zone

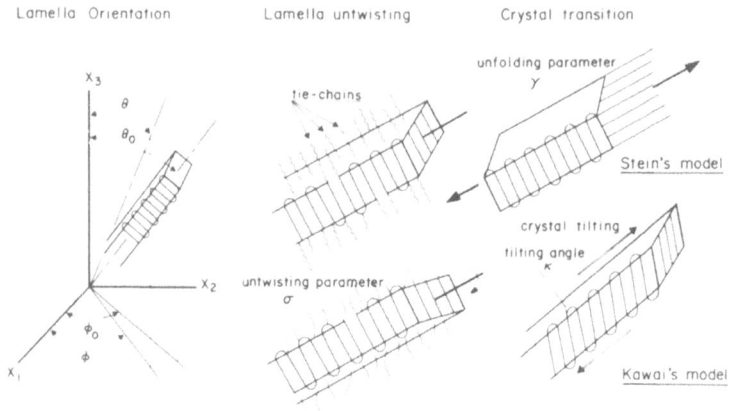

FIG. 6. Schematic diagrams illustrating the deformation mechanism of poly-ethylene spherulites under uniaxial stretching. Upper: Stein's model taking account of crystal transition and lamellar untwisting; lower: Kawai's model taking account of crystal tilting and lamellar untwisting.

(the highest shear stress zone) of the spherulite in a contrast fashion represented by a preferential rotation of the crystal grain around its own a axis so as to make the c axis parallel to the lamellar axis, i.e. lamellar tilting associated with micronecking of the lamella and/or inter-lamellar shearing also due to straining of the tie-chain molecules between adjacent lamellae, as schematised in Fig. 6.

Finally, it should be noted that the orientation distribution function, $w(\cos\theta, \phi, \eta)$ thus determined is much more straightforward for describing the state of orientation or orientation behaviour of the crystal grains (crystallites) than an assembly of the same number of the so-called pole-figures with j stereo-projections of $q_j(\zeta_j, \phi_j)$ on either plane perpendicular to the $x_1, x_2,$ or x_3 axis, though $w(\cos\theta, \phi, \eta)$ thus determined is the lth order approximation based on a finite series of the spherical harmonics up to the lth order rather than the infinite series.

3. ESTIMATION OF ORIENTATION DISTRIBUTION FUNCTION FROM FINITE SERIES OF SPHERICAL HARMONICS, AND TRUNCATED ERROR

As was mentioned previously, the orientation distribution functions, either $w(\xi, \phi, \eta)$ or $q_j(\zeta_j, \phi_j)$, for the structural elements are not necessarily determined from experimental sources, except for crystalline elements for which the X-ray diffraction measurements give $q_j(\zeta_j, \phi_j)$ and, subsequently, $w(\xi, \phi, \eta)$, as has been discussed just before, with the lth order approximation. In general, the orientation of noncrystalline structural elements is obtained as a limited number of its averages, i.e. the second and/or fourth moments of the orientation distribution function $q_j(\zeta_j, \phi_j)$ and further the corresponding orientation factors f_{lm}^j, from measurements of optical anisotropy of the material, such as birefringence and absorption and emission dichroisms, as will be discussed in detail in the following section. It is, therefore, emphasised that the above aspects give a serious problem in understanding the bulk properties of the polymeric materials, irrespective of crystalline or noncrystalline materials, which are, more or less, related to the orientation of the noncrystalline structural elements.

Let us discuss the estimation of the orientation distribution function, $q_j(\zeta_j, \phi_j)$ from the limited number of the coefficients of the expanded series in the spherical harmonics, i.e. the quantities related to the measurable orientation factors, as just mentioned above, at least up to the fourth order, and a truncated error of the estimation of the real distribution function, quantitatively. For simplicity, the discussion will be limited to a system of uniaxial symmetry.[91] It can, however, be extended without difficulty to a system of orthogonal–biaxial symmetry.[89, 90]

The mean-square error between the real distribution function and the truncated function from a finite series of the expansion with $l \leq l_1$ is given by

$$\sigma_{l_1} = \int_0^{2\pi} \int_{-1}^{1} [q_j(\zeta_j, 0) - \sum_{l=0}^{l_1} \alpha_{l0}^j \Pi_l^0(\zeta_j)]^2 \, d\zeta_j \, d\phi_j$$

$$= 2\pi \int_{-1}^{1} [q_j(\zeta_j, 0) - \sum_{l=0}^{l_1} \alpha_{l0}^j \Pi_l^0(\zeta_j)]^2 \, d\zeta_j \quad (38)$$

where l must be an even number. From the orthogonal properties of the Legendre polynomials, eqn (38) may be reduced to

$$\sigma_{l_1} = 2\pi \left[\int_{-1}^{1} \{q_j(\zeta_j, 0)\}^2 \, d\zeta_j - \sum_{l=0}^{l_1} (\alpha_{l0}^j)^2 \right] \quad (39)$$

For random orientation, $q_j(\zeta_j, 0) = (1/4\pi)$, and $\alpha_{l0}^j = 1/(2\pi\sqrt{2})$ for $l = 0$ or $\alpha_{l0}^j = 0$ for $l \neq 0$. This means that the mean-square error is zero for random orientation, even when l is limited as small as zero-th order. On the other hand, the sharper the distribution function, the larger the value within the integration in the right-hand side of eqn (39) and the error become, unless l_1 is taken as counter-balancingly high. This means that the sharper the orientation distribution, the higher the order of l_1 that should be taken into account to approach the real distribution.

Let us evaluate the error quantitatively by using a uniaxial orientation distribution function proposed by Kratky[16] for orientation of floating rods in an affine matrix; i.e.

$$q_j(\zeta_j) = (1/4\pi)\lambda^3/[\lambda^3 - (\lambda^3 - 1)\zeta_j^2]^{3/2} \tag{40}$$

where λ is the uniaxial extension ratio of the bulk sample.

Figure 7 shows the change of $\int_{-1}^{1}[q_j(\zeta_j)]^2 \, d\zeta_j$ and of the error for various values of l_1 up to 30, both with increase of f_{20}^j accompanied with increase of λ. As clearly seen in the figure, it can be pointed out that the larger the value of f_{20}^j associated with the sharpening of the orientation distribution as

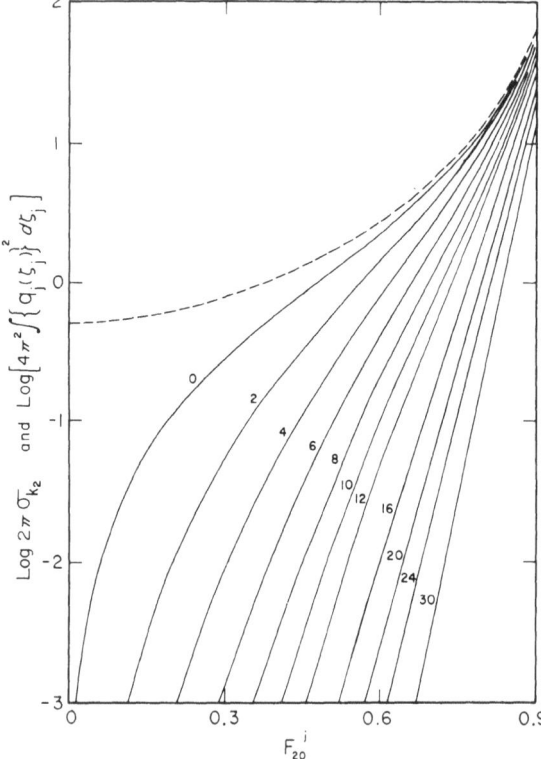

FIG. 7.   Changes of $\int_{-1}^{1}[q_j(\zeta_j)]^2 \, d\zeta_j$ (broken curve) and of the truncated error (solid curves) for various values of l_1 up to 30, both with increase of f_{20}^j accompanied with increase of λ in Kratky's function of eqn (40).

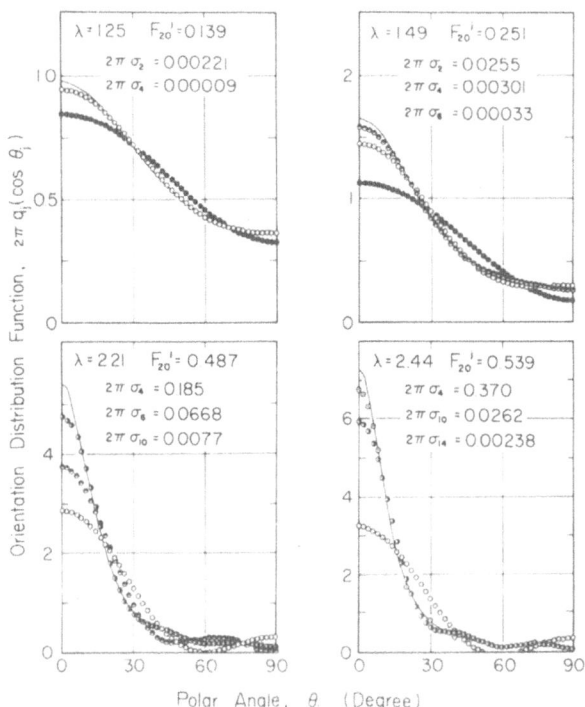

FIG. 8. Comparison of Kratky's orientation distribution functions at various extension ratios (solid curves) with the truncated functions; (●) $l_1 = 2$; (○) $l_1 = 4$; (◕) $l_1 = 6$; (◑) $l_1 = 10$; and (⊙) $l_1 = 14$.

λ increases, the higher the value of l_1 needed to keep the error within a given level.

Figure 8 shows a comparison of Kratky's distribution function at various extension ratios, which is indicated by a solid line, with the truncated function calculated from the finite series of the expansion. It can be seen again that the larger the extension ratio, i.e. the larger the value of f_{20}^i, the higher the value of l_1 necessary to approach the real distribution function.

4. OPTICAL ANISOTROPY AND ORIENTATION OF STRUCTURAL ELEMENTS

Optical anisotropy of polymeric materials in bulk must be ascribed to the orientation of certain structural elements which are also optically anisotropic. Mathematically, the optical anisotropy in bulk can be represented in terms of the moments of the orientation distributions of the structural elements and their intrinsic optical constants. The structural elements for semi-crystalline polymers may be taken as the crystallites and non-crystalline chain segments on the basis of a two-phase hypothesis of crystalline and non-crystalline regions. Therefore, the moments of orientation distributions of the elements are important parameters not only for characterising the orientation behaviour of bulk polymers during processing, but also for predicting the anisotropy of physical properties, such as mechanical and optical properties, of the processed materials. In this section, some indices of the optical anisotropy in bulk, such as birefringence and absorption and/or emission dichroisms, will be discussed in a general manner based on the optical dichroic ratios which are, in turn, related to the generalised orientation factors defined previously.

4.1. General Description of Optical Absorption and/or Emission Behaviour

The polymeric materials, in which optically anistropic structural elements preferentially orient, result in both anisotropic absorption and emission of rays to some extent depending on the wavelength irradiating the materials. The principle of fluorescence polarisation has been discussed by a number of authors[69–77] in somewhat different ways, which may be generalised here. Let us consider a *chromophoric group* excited by polarised radiation, where the absorption and emission anisotropies of the group are assumed for simplicity as being rotational-cylindrical symmetry around an identical axis fixed within the fluorescent which is further fixed within the structural element, as illustrated in Fig. 9. The observed fluorescence intensity from the bulk material can be given by a product of the absorption and emission anisotropic tensors (second rank) of the material as follows:

$$I(\mathbf{P},\mathbf{Q}) = KP_i A_{ij} P_j Q_p E_{pq} Q_q \qquad (41)^*$$

where K is an instrumental constant, and $\mathbf{P}$ and $\mathbf{Q}$ are polarisation directions of the polariser for the incident ray and of the analyser for the emitted ray, respectively. P_i and Q_p are direction cosines of the polarisation directions of the polariser and analyser, and A_{ij} and E_{pq} are second rank tensors for the anisotropic absorption and emission of the bulk material, all with respect to the laboratory coordinates $0-x_1 x_2 x_3$ fixed within the bulk material, as shown in Fig. 10.

---

* Hereafter the asterisk indicates the summation convention.

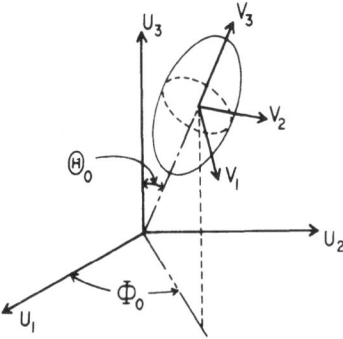

FIG. 9. Schematic diagram illustrating orientation of the chromophoric group of absorption and emission anisotropies with respect to the fluorescent and further to the structural element of $0-u_1 u_2 u_3$.

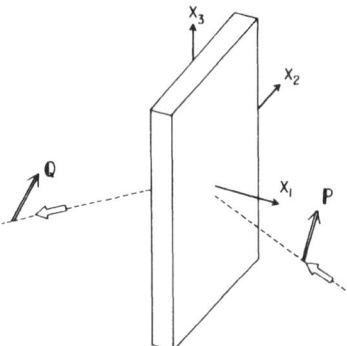

FIG. 10. Schematic diagram showing the geometry of the optical system for fluorescence polarisation measurements with respect to the bulk sample of $0-x_1 x_2 x_3$.

The rotational–cylindrical symmetry of absorption and emission aniso-tropies of the chromophoric group can be represented also in terms of the second rank tensors as follows:

$$(a_{kl}^0) = \begin{pmatrix} a_{11}^0 & 0 & 0 \\ 0 & a_{11}^0 & 0 \\ 0 & 0 & a_{33}^0 \end{pmatrix} \tag{42}$$

and

$$(e^0_{rs}) = \begin{pmatrix} e^0_{11} & 0 & 0 \\ 0 & e^0_{11} & 0 \\ 0 & 0 & e^0_{33} \end{pmatrix} \qquad (43)$$

where (a^0_{kl}) and (e^0_{rs}) can be related to (A_{ij}) and (E_{pq}), respectively, with the transformation of the second rank tensors as follows:

$$A_{ij} = t_{ik}t_{jk}a^0_{kk} \qquad (44)^*$$

$$E_{pq} = t_{pr}t_{qr}e^0_{rr} \qquad (45)^*$$

where, for example, t_{ik} is the direction cosine between the x_i axis fixed within the bulk material and the v_k axis fixed at the chromophoric group. Then, the fluorescence intensity from the bulk material can be written by averaging over all orientations of the chromophoric groups as follows:

$$I(\mathbf{P},\mathbf{Q}) = KP_iP_jQ_pQ_q \langle t_{ik}t_{jk}t_{pl}t_{ql} \rangle a^0_{kk}e^0_{ll} \qquad (46)^*$$

$$= KP_iP_jQ_pQ_q \langle F_{ijpq} \rangle \qquad (47)^*$$

With the orthogonal–biaxial symmetry of the orientation distribution of the structural elements, the components, $\langle F_{ijpq} \rangle$, in eqn (47) may be written as:[73]

$$L_{ij} = \langle F_{iijj} \rangle /(a_0 e_0) = [\langle t^2_{i3}t^2_{j3} \rangle (a^0_{33} - a^0_{11})(e^0_{33} - e^0_{11})$$
$$+ \langle t^2_{i3} \rangle (a^0_{33} - a^0_{11})e^0_{11}$$
$$+ \langle t^2_{j3} \rangle (e^0_{33} - e^0_{11})a^0_{11} + a^0_{11}e^0_{11}]/(a_0 e_0) \qquad (48)$$

$$L'_{ij} = \langle F_{ijij} \rangle /(a_0 e_0) = \langle t^2_{i3}t^2_{j3} \rangle (a^0_{33} - a^0_{11})(e^0_{33} - e^0_{11})/(a_0 e_0) \qquad (49)$$
$$L_{ijkl} = \langle F_{ijkl} \rangle /(a_0 e_0) = 0 \qquad (50)$$

where $a_0 = a^0_{33} + 2a^0_{11}$, $e_0 = e^0_{33} + 2e^0_{11}$, and $\langle F_{iijj} \rangle \neq \langle F_{jjii} \rangle \neq \langle F_{ijij} \rangle$

Equations (47) through (49) show that the fluorescence intensity for all possible arrangements of $\mathbf{P}$ and $\mathbf{Q}$ can be determined by twelve components of $\langle F_{iijj} \rangle$ and $\langle F_{ijij} \rangle$. However, as can be seen in the right-hand sides of eqns (48) and (49), unknown parameters are two intrinsic optical constants, $(a^0_{33} - a^0_{11})/a_0$ and $(e^0_{33} - e^0_{11})/e_0$, and five moments of $\langle t^2_{i3}t^2_{j3} \rangle$, but not nine moments, because of some of the nine moments being related to the linear combinations of $\langle t^2_{i3}t^2_{j3} \rangle$.

At a given geometry of the optical system with respect to the bulk sample, one may define the dichroic ratio as

$$D_k(\mathbf{P},\mathbf{Q}) = I(\mathbf{P},\mathbf{Q})/I(\mathbf{x}_3,\mathbf{x}_3) \tag{51}$$

where the instrumental constant is cancelled out and the error is minimised. For the incident radiation perpendicular or inclined at $\pi/4$ to the surface of a bulk sample (the x_2x_3 plane), the dichroic ratios can be related to simple combinations of L_{ij}, which should be determined at first, as follows:[113]

$$D_1(\mathbf{x}_2,\mathbf{x}_2) = L_{22}/L_{33}$$

$$D_2(\mathbf{x}_3,\mathbf{x}_2) = L_{32}/L_{33}$$

$$D_3(\mathbf{x}_2,\mathbf{x}_3) = L_{23}/L_{33} \tag{52a}$$

$$D_4(\mathbf{x}_3, \cos(\pi/4)\mathbf{x}_1 + \sin(\pi/4)\mathbf{x}_2) = (1/2)(L_{31} + L_{32})/L_{33}$$

$$D_5(\mathbf{x}_2, \cos(\pi/4)\mathbf{x}_1 + \sin(\pi/4)\mathbf{x}_2) = (1/2)(L_{21} + L_{22})/L_{33} \tag{52b}$$

$$D_6(-\cos(\pi/4)\mathbf{x}_1 + \sin(\pi/4)\mathbf{x}_2, \mathbf{x}_3) = (1/2)(L_{13} + L_{23})/L_{33}$$

$$D_7(-\cos(\pi/4)\mathbf{x}_1 + \sin(\pi/4)\mathbf{x}_2, \mathbf{x}_2) = (1/2)(L_{12} + L_{22})/L_{33} \tag{52c}$$

$$D_8(-\cos(\pi/4)\mathbf{x}_1 + \sin(\pi/4)\mathbf{x}_2, \cos(\pi/4)\mathbf{x}_1 + \sin(\pi/4)\mathbf{x}_2)$$

$$= (1/4)(L_{11} + L_{12} + L_{21} + L_{22} - 4L'_{12})/L_{33}$$

$$D_9(\cos(\pi/4)\mathbf{x}_1 + \sin(\pi/4)\mathbf{x}_2, \cos(\pi/4)\mathbf{x}_1 + \sin(\pi/4)\mathbf{x}_2)$$

$$= (1/4)(L_{11} + L_{12} + L_{21} + L_{22} + 4L'_{12})/L_{33} \tag{52d}$$

where eqns (52a,b) are the results for the perpendicular radiation, while eqns (52c,d) are those for the inclined radiation for which the refraction of the incident ray at the surface of the bulk sample must be avoided by use of immersion[72] or hemisphere[98] technique, or corrected appropriately.

Using a combination of the dichroic ratios given by

$$D_0 = 2(D_4 + D_6 + D_8 + D_9) + 1 = \sum_{i=1}^{3} \sum_{j=1}^{3} L_{ij}/L_{33} = 1/L_{33} \tag{53}$$

the component L_{ij} can be obtained as follows:

$$L_{11} = (2D_8 + 2D_9 - 2D_7 - 2D_5 + D_1)/D_0$$

$$L_{12} = (2D_7 - D_1)/D_0$$

$$L_{13} = (2D_6 - D_3)/D_0$$

$$L_{21} = (2D_5 - D_1)/D_0$$

$$L_{22} = D_1/D_0$$

$$L_{23} = D_3/D_0$$

$$L_{31} = (2D_4 - D_2)/D_0$$

$$L_{32} = D_2/D_0$$

$$L_{33} = 1/D_0 \qquad (54)$$

In the fluorescence polarisation technique, the intrinsic optical constants, $(a_{33}^0 - a_{11}^0)/a_0$ and $(e_{33}^0 - e_{11}^0)/e_0$, can be determined. With random orientation, one may obtain the following relation from eqns (48) and (52):

$$\alpha_2 = (5/2)\frac{1 - (D_2 + D_3)/2}{1 + (D_2 + D_3)} = (5/2)\frac{L_{33} - (L_{23} + L_{32})/2}{L_{33} + L_{23} + L_{32}}$$

$$= \frac{(a_{33}^0 - a_{11}^0)(e_{33}^0 - e_{11}^0)}{a_0 e_0} \qquad (55)$$

where the value of α_2 varies from zero to unity, depending on the absorption and emission anisotropies of the chromophoric group, i.e. α_2 being unity for the extreme anisotropies of a_{11}^0 and $e_{11}^0 = 0$, and being zero for isotropies of $a_{11}^0 = a_{33}^0$ and $e_{11}^0 = e_{33}^0$. On the other hand, one can obtain the material constant independent of the degree of orientation as follows:

$$\alpha_1 = \frac{3(D_5 - D_7)}{D_8 + D_9 + D_6 - D_3 - 2D_5} + 1 = \frac{3(L_{21} - L_{12})}{\sum_{j=1}^{3}(L_{1j} - L_{2j})} + 1$$

$$= \frac{a_0(e_{33}^0 - e_{11}^0)}{(a_{33}^0 - a_{11}^0)e_0} \qquad (56)$$

Here α_1 is unity for the extreme anisotropies, i.e. $L_{ij} = L_{ji}$. Then, the intrinsic optical constants of absorption and emission anisotropies can be determined as follows:

$$\begin{aligned}(a_{33}^0 - a_{11}^0)/a_0 &= (\alpha_2/\alpha_1)^{1/2}\\ (e_{33}^0 - e_{11}^0)/e_0 &= (\alpha_2\alpha_1)^{1/2}\end{aligned} \qquad (57)$$

Further, one can calculate the moments of $\langle t_{i3}^2 t_{j3}^2 \rangle$ from the linear combinations of L_{ij} and the intrinsic optical constants. As a result, the fluorescence polarisation has an additional advantage over ordinary absorption or emission dichroisms, in which the values of the intrinsic optical and material constants cannot be obtained.

For the fluorescence polarisation, on the other hand, simple relations of the measurable dichroic ratios and/or the components of L_{ij} to the second and fourth order orientation factors cannot be obtained. Thus, the moments, $M_{ij} = \langle t_{i3}^2 t_{j3}^2 \rangle$, which are derived from L_{ij} and also from the dichroic ratios D_k, may be related to the second and fourth order orientation factors as

$$f_{20}^0 = (3 \sum_{i=1}^{3} M_{i3} - 1)/2 = (3 \sum_{j=1}^{3} M_{3j} - 1)/2$$

$$f_{22}^0 = 3 \sum_{i=1}^{3} (M_{i1} - M_{i2}) = 3 \sum_{j=1}^{3} (M_{1j} - M_{2j}) \tag{58}$$

$$f_{40}^0 = (35 M_{33} - 30 \sum_{i=1}^{3} M_{i3} + 3)/8$$

$$f_{42}^0 = 15[7(M_{31} - M_{32}) - \sum_{i=1}^{3} (M_{i1} - M_{i2})]/2$$

$$f_{44}^0 = 105(M_{11} + M_{22} - 6M_{12}) \tag{59}$$

Similarly to the derivation of eqn (11), the orientation factors f_{lmn} of the structural element, possibly of a non-crystalline element involving the fluorescent, may be related to the above orientation factors as follows:

$$f_{lm}^0 = P_l(\cos \Theta_0) f_{lm0} + 2 \sum_{n=2}^{l} \frac{(l-n)!}{(l+n)!} P_l^n(\cos \Theta_0) \cos(n\Phi^0) f_{lmn} \tag{60}$$

where Θ_0 and Φ_0 are the polar and azimuthal angles specifying the orientation of the chromophoric group with respect to the structural element, as shown in Fig. 9, and $l(\leq 4)$, $m(\leq l)$, and n are even numbers. In most of the non-crystalline orientation, it may be assumed that the orientation of the structural elements, the non-crystalline chain segments, is random around their own u_3 axes, i.e. random orientation with respect to the rotational angle η. Thus eqn (60) can be rewritten as

$$f_{lm}^0 = P_l(\cos \Theta_0) f_{lm0} \tag{61}$$

where $l = 2$ or 4, and $m(\leq l) = 0$, 2 or 4.

The absorption or emission dichroisms may be considered as a special case of the fluorescence polarisation. Under the polarisation radiation, **P**, the intensity absorbed by the bulk material may be written from eqns (41) through (47) as

$$A(\mathbf{P}) = KP_iP_j \sum_{k=1}^{3} \langle F_{ij;kk} \rangle / e_0 = KP_iP_j \langle A_{ij} \rangle$$

$$= KP_iP_j \langle t_{ik}t_{jk} \rangle a_{kk}^0 \qquad (62)^*$$

Similarly, if an unpolarised incident ray perpendicular to the x_2x_3 plane is used, the observed emission intensity through the analyser, $\mathbf{Q}$, is given by

$$E(\mathbf{Q}) = KQ_pQ_q \sum_{i=2}^{3} \langle F_{iipq} \rangle \qquad (63)^*$$

However, the electric vector of the incident ray is parallel to the x_2x_3 plane, and any simple relation cannot be obtained.

For the orthogonal–biaxial orientation, the components $\langle A_{ii} \rangle$ in eqn (62) may be rewritten as

$$\langle A_{ii} \rangle = \langle t_{i3}^2 \rangle (a_{33}^0 - a_{11}^0) + a_{11}^0 \qquad (64)$$

That is, the absorption dichroism is expressed as a special case of the fluorescence polarisation.

The dichroic ratios for the absorption dichroism are also defined as

$$D_k(\mathbf{P}) = A(\mathbf{P})/A(\mathbf{x}_3) \qquad (65)$$

Then, the dichroic ratios for the perpendicular and inclined radiations can be rewritten, respectively, as

$$D_1(\mathbf{x}_2) = \frac{\langle t_{23}^2 \rangle (a_{33}^0 - a_{11}^0) + a_{11}^0}{\langle t_{33}^2 \rangle (a_{33}^0 - a_{11}^0) + a_{11}^0} \qquad (66)$$

$$D_2(-\cos(\pi/4)\mathbf{x}_1 + \sin(\pi/4)\mathbf{x}_2) =$$
$$\frac{(\langle t_{13}^2 \rangle \times \langle t_{23}^2 \rangle)(a_{33}^0 - a_{11}^0)/2 + a_{11}^0}{\langle t_{33}^2 \rangle (a_{33}^0 - a_{11}^0) + a_{11}^0} \qquad (67)$$

The dichroic orientation factors may be defined as

$$F_D^{20} = (1 - D_2)/(1 + 2D_2) \qquad (68)$$

$$F_D^{22} = (D_2 - D_1)/(1 + 2D_2) \qquad (69)$$

Substituting the results of eqns (66) and (67) into eqns (68) and (69), it follows that

$$F_D^{20} = \frac{a_{33}^0 - a_{11}^0}{a_{33}^0 + 2a_{11}^0} \frac{3\langle t_{33}^2 \rangle - 1}{2} = \frac{a_{33}^0 - a_{11}^0}{a_{33}^0 + 2a_{11}^0} f_{20}^0 \qquad (70)$$

$$F_D^{22} = \frac{a_{33}^0 - a_{11}^0}{a_{33}^0 + 2a_{11}^0} \frac{\langle t_{13}^2 \rangle - \langle t_{23}^2 \rangle}{2} = \frac{a_{33}^0 - a_{11}^0}{a_{33}^0 + 2a_{11}^0} \frac{f_{22}^0}{6} \qquad (71)$$

As can be seen from eqns (70) and (71), the dichroic orientation factors are represented by the product of the orientation of the chromophoric group and its intrinsic optical constants. The constants, however, cannot be obtained from the absorption dichroism alone and must be determined from other sources. When the orientation factors, f_{20}^0 and f_{22}^0, are obtained from eqns (70) and (71), eqns (60) and (61) also hold for $l = 2$, and the orientation factors of the structural elements f_{2mn} may be determined, providing that the polar and azimuthal angles, Θ_0 and Φ_0, are known also from other sources.

4.2. Birefringence

Birefringence, anisotropy of refractive index in a bulk material, is observed as a sum of the form, distortional, and orientational birefringences. In most cases, the former two contributions to the total bulk birefringence are neglected because of their small magnitudes and will not be discussed here. For the orthogonal–biaxially oriented materials, three principal refractive indices, N_1, N_2, and N_3, are different from each other and lie along the normal direction (x_1 axis), the transverse direction (x_2 axis), and the machine direction (x_3 axis) of the bulk material, respectively. Then, three kinds of birefringence may be defined as

$$\Delta_{31} = N_3 - N_1$$

$$\Delta_{32} = N_3 - N_2 \qquad (72)$$

$$\Delta_{21} = N_2 - N_1$$

where one of them is linearly dependent on the other two, i.e. $\Delta_{31} = \Delta_{32} + \Delta_{21}$. One can measure Δ_{32} the most easily for a sheet-like sample under perpendicular radiation to the sheet surface, and Δ_{21} with the tilted technique proposed by Stein.[114] For the uniaxially oriented material, only one of them, Δ_{32}, is sufficient, because $\Delta_{32} = \Delta_{31}$ and $\Delta_{21} = 0$.

For an isotropic system, the refractive index n can be related to polarisability per unit volume P by the Lorentz–Lorenz formula as

$$\frac{n^2 - 1}{n^2 + 2} = \frac{4\pi}{3} P \tag{73}$$

Applying eqn (73) to an anisotropic system and differentiating both sides of the formula, the small difference in refractive indices, i.e. the bire-fringence, can be related to the polarisability difference as follows:

$$\Delta_{ij} = N_i - N_j = \frac{2\pi (\bar{n}^2 + 2)^2}{9} \frac{}{\bar{n}} (P_{ii} - P_{jj}) \tag{74}$$

where $\bar{n}$ is the mean refractive index of the anisotropic system, and the polarisability P_{ii} is a tensor quantity with a second rank and may be discussed in the same manner as eqns (62) through (71) just by replacing $\langle A_{ii} \rangle$ with P_{ii}. Then, the polarisability of the bulk material may be given by

$$P_{ii} = \langle t_{ij}^2 \rangle \alpha_{jj} \tag{75}*$$

where α_{jj} is the principal polarisability of the structural element of ortho-rhombic symmetry.

Finally, the relationships between the birefringence and the orientation factors of the structural elements may be derived as follows:[61]

$$\frac{(\Delta_{31} + \Delta_{32})}{2} = \frac{2\pi}{9} [P_{33} - (P_{11} + P_{22})/2]$$

$$= \frac{2\pi (\bar{n}^2 + 2)^2}{9} \frac{}{\bar{n}} \left[\left(\alpha_{33} - \frac{\alpha_{11} + \alpha_{22}}{2} \right) f_{200} + (\alpha_{11} - \alpha_{22}) f_{202}/4 \right] \tag{76}$$

$$= \frac{2\pi (\bar{n}^2 + 2)^2}{9} \frac{}{\bar{n}} [(\alpha_{11} - \alpha_{33}) f_{20}^1 + (\alpha_{22} - \alpha_{33}) f_{20}^2] \tag{76a}$$

$$\Delta_{32} - \Delta_{31} = \frac{2\pi (\bar{n}^2 + 2)^2}{9} \frac{}{\bar{n}} (P_{11} - P_{22})$$

$$= \frac{2\pi (\bar{n}^2 + 2)^2}{9} \frac{}{\bar{n}} \left[\left(\alpha_{33} - \frac{\alpha_{11} + \alpha_{22}}{2} \right) f_{220} + (\alpha_{11} - \alpha_{22}) f_{222}/4 \right]/3 \tag{77}$$

$$= \frac{2\pi (\bar{n}^2 + 2)^2}{9} \frac{}{\bar{n}} [(\alpha_{11} - \alpha_{33}) f_{22}^1 + (\alpha_{22} - \alpha_{33}) f_{22}^2]/3 \tag{77a}$$

Applying eqn (74) to the intrinsic birefringence of the structural element, eqns (76a) and (77a) may be expressed only in terms of refractive indices as follows:

$$N_3 - (N_1 + N_2)/2 = (n_1^0 - n_3^0) f_{20}^1 + (n_2^0 - n_3^0) f_{20}^2 \tag{78}$$

$$N_1 - N_2 = [(n_1^0 - n_3^0) f_{22}^1 + (n_2^0 - n_3^0) f_{22}^2]/3 \tag{79}$$

where n_i^0 is the principal refractive index of the orthorhombic structural element. Equation (78) is the same equation as derived by Stein[99] for the orientation of orthorhombic polyethylene crystals.

For a cylindrical structural element which is transversely isotropic, such as the non-crystalline chain segment, the relations of eqns (76) and (77) may be reduced to

$$N_3 - (N_1 + N_2)/2 = \frac{2\pi (\bar{n}^2 + 2)^2}{9} \frac{}{\bar{n}} (\alpha_{33} - \alpha_{11}) f_{200} \tag{80}$$

$$N_1 - N_2 = \frac{2\pi (\bar{n}^2 + 2)^2}{9} \frac{}{\bar{n}} (\alpha_{33} - \alpha_{11}) f_{220}/3 \tag{81}$$

which correspond to eqns (70) and (71), respectively.

In the case of a uniaxially stretched rubber network having Gaussian chains, the birefringence has been calculated by Kuhn and Grün[44] and Treloar[45] as

$$N_3 - N_2 = \frac{2\pi (\bar{n}^2 + 2)^2}{9} \frac{\nu}{\bar{n}} \frac{}{5} (\alpha_{33} - \alpha_{11}) (\lambda^2 - 1/\lambda) \tag{82}$$

where ν is the number of network chains per unit volume and λ is the extension ratio of the system.

The birefringence of a multiphase system, such as semi-crystalline polymers and polymer blends of incompatible components, may be considered as the sum of contributions from respective phases and is given by

$$\Delta_{\text{total}} = \sum_i X_i \Delta_i + \Delta_f \tag{83}$$

where Δ_{total}, Δ_f, and Δ_i are total birefringence of the system, form birefringence, and orientational birefringence of the ith phase, respectively, and X_i is the volume fraction of the ith phase. One can obtain the orientation factors of the crystalline phase from X-ray diffraction and the birefringence of the crystalline contribution from eqns (78) and (79). In turn, neglecting the form birefringence and subtracting the crystalline contribution from the total birefringence, one can determine the non-crystalline contribution and further the orientation factors of the non-crystalline chain segments from eqns (80) and (81), providing that the intrinsic birefringences of the structural elements are available.

4.3. Absorption Dichroism

Conjugated double bonds, polyene segments, or dye molecules absorbed in a polymeric material give the anisotropic absorption for the ultraviolet (UV) and visible polarised monochromatic radiations. For the infrared (IR) region, various kinds of bonds and groups, such as N—H, C=O, and CH_2, can give rise to the anisotropic absorption at their specific wavelengths. In most cases, the anisotropy has been treated as being perfect such that, for example, a C=O bond exhibits a maximum absorption when the electric vector of the polarised radiation is parallel to the bond but no absorption at the perpendicular polarisation. A problem in the perfect and imperfect anisotropies will be discussed later.

The absorbance defined according to the Lambert–Beer law is given by

$$A = \log(I_0/I) \qquad (84)$$

where I_0 and I are the intensities of the incident and transmitted rays, respectively. The absorbance varies depending on the wavelength and shows a maximum at a characteristic wavelength related to a specific chemical structure of the absorbent. Close to the characteristic wavelength for the specific chemical structure, the anisotropic absorption may be seen and be proportional to the square of the direction cosine between the transition moment for the specific chemical structure and the electric vector of the polarised radiation. That is, the absorbance of the bulk material is given by eqn (62), and the dichroic ratios and the dichroic orientation factors of the transition moment may be defined similarly as eqns (65) through (69), where the imperfect anisotropy of making $a_{11}^0 \neq 0$ is taken in account.

Substitution of eqn (61) into eqns (70) and (71) results in

$$f_D^{2m} = C_m \frac{a_{33}^0 - a_{11}^0}{a_{33}^0 + 2a_{11}^0} \frac{3\cos^2\Theta_0 - 1}{2} f_{2m0} \qquad (85)$$

where C_m is 1 or 1/6 for $m = 0$ or 2, respectively. As can be seen in eqn (85), the dichroic orientation factor f_D^{2m} to be observable is represented by a product of the intrinsic optical anisotropy of the absorbant (a_{11}^0/a_{33}^0), the orientation of the absorbant to the structural element (Θ_0), and the orientation of the structural element with respect to the bulk material (f_{2m0}). It is, however, uncertain whether a_{11}^0 is negligibly small and (a_{11}^0/a_{33}^0)

and (Θ_0) can be determined separately from other sources. Therefore, it may be convenient to define a material constant as

$$G = \frac{a_{33}^0 + 2a_{11}^0}{a_{33}^0 - a_{11}^0} \frac{2}{3\cos^2\Theta_0 - 1} \qquad (86)$$

then the orientation factors of the structural element can be obtained as

$$f_{2m0} = Gf_D^{2m}/C_m \qquad (87)$$

In the IR dichroism and the UV or visible dichroism of polyene segments, the absorbant must be oriented at a given angle within the structural element. However, in the dye dichroism, one cannot assure that the dye molecules are oriented with a fixed angle irrespective of orientation of the polymer chains.

4.4. Fluorescence and Raman Polarisations

Both cases of fluorescence and Raman polarisations can be discussed in an identical manner, but the chromophoric groups and, therefore, the wavelengths to be employed are different. The chromophoric groups are excited by the polarised radiations and the polarisation of the emitted rays is observed. Mathematical derivation of the relationship between the polarisation of the emitted ray and the orientation of the chromophoric group has already been discussed in section 4.1. One can obtain the second and fourth order orientation factors, f_{lm}^0, where $l = 4$ and 2, and $l \geq m = 4$, 2 and 0. f_{lm}^0 may be related to the orientation factor of the structural element as

$$f_{lm}^0 = P_l(\cos\Theta_0) f_{lm0} \qquad (88)$$

where Θ_0 is the orientation angle of the chromophoric group with respect to the structural element and may be determined for the Raman polarisation with the aid of the chemical structure of the polymer chain, while its determination for the fluorescence polarisation has the same problem as discussed above for the dye dichroism.

4.5. Anisotropy in NMR

In a material composed of one kind of magnetic nuclei, the second moment of the NMR absorption line has been derived by Van Vleck[80] as follows:

$$\langle \Delta H^2 \rangle = \frac{G}{N} \sum_{j > k} \left\langle \left(\frac{3\cos^2\beta_{jk} - 1}{r_{jk}^3} \right)^2 \right\rangle \qquad (89)$$

where G is a constant related to the nature of the spin, N the total number

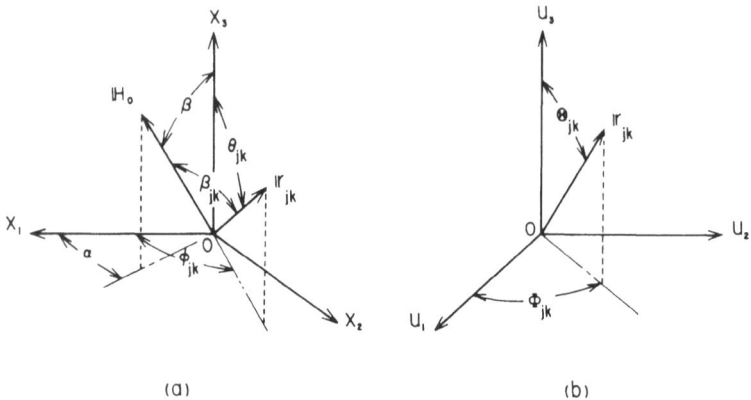

(a)   (b)

FIG. 11. Diagrams illustrating the orientation of the distance vector r_{jk} and the magnetic field H_0 both with respect to: (a) the bulk sample of $0-x_1x_2x_3$; and (b) the structural element of $0-u_1u_2u_3$.

of nuclei summed up, r_{jk} is the distance between the jth and kth nuclei, and β_{jk} the angle between the vector r_{jk} and the applied magnetic field H_0, as illustrated in Fig. 11, in which the orientations of r_{jk} and H_0 with respect to the bulk material $0-x_1x_2x_3$ and the structural element $0-u_1u_2u_3$ are defined.

Anisotropy of the second moment in the rigid lattice, where the molecular motion is frozen, for example, at liquid nitrogen temperature, may be represented in terms of the orthogonal–biaxial orientation of the orthorhombic elements as[74, 80-82, 115]

$$\langle \Delta H^2 \rangle = 4 \sum_{l=0}^{4} \sum_{m=-l}^{l} R_{lm} \Pi_l^m (\cos \beta) \cos (m\alpha) \tag{90}$$

where a set of angles (β, α) is the alignment angle of H_0, as shown in Fig. 11, and R_{lm} is a material constant depending on the orientation and structure of the element as given by

$$R_{lm} = R_{l\bar{m}} = \left(\frac{2}{2l+1} \right)^{1/2} \sum_{n=-l}^{l} S_{ln} A_{lmn} \tag{91}$$

and where S_{ln} is the so-called lattice sum and is given by

$$S_{ln} = S_{l\bar{n}} = \frac{G}{N} a_l \left(\frac{2}{2l+1} \right)^{1/2} \sum_{j>k} \frac{\Pi_l^n (\cos \Theta_{jk}) \cos (n\Phi_{jk})}{r_{jk}^6} \tag{92}$$

Here $a_0 = \sqrt{2}/5$, $a_2 = 2\sqrt{10}/35$, $a_4 = 6\sqrt{2}/35$, and A_{lmn} is one of the coefficients of the expanded series, as defined by eqn (11). At six or more alignments of H_0, one can obtain the independent values of R_{lm}. However, as can be recognised from eqn (91), one cannot calculate A_{lmn} for the orthorhombic element without any other sources of experiments nor any assumption on the orientation distribution of the element with respect to the rotational angle η.[81] For the transversely isotropic element for which $S_{ln} = A_{lmn} = 0$ for $n \neq 0$, eqn (91) may be reduced as

$$R_{lm} = \left(\frac{2}{2l+1}\right)^{1/2} S_{l0} A_{lm0} \tag{93}$$

Then, one can determine the average values of the orientation distribution of the elements by using the theoretical values of the lattice sum S_{l0}.

Under circumstances under which the molecular motions are activated, one must take averages not only over the line width splitting due to possible kinds of molecular motions, but also over all of the orientations of the vector r_{jk}. That is, the averagings may be written as

$$\left\langle \frac{3\cos^2\beta_{jk} - 1}{r_{jk}^3} \right\rangle_{motion} = \frac{4}{5}[A\Pi_2^0(\cos\beta') + 2B\Pi_2^2(\cos\beta')\cos 2\alpha'] \tag{94}$$

where

$$A = \left\langle \frac{\Pi_2^0(\cos\Theta_{jk})}{r_{jk}^3} \right\rangle_{motion} \tag{95}$$

$$B = \left\langle \frac{\Pi_2^2(\cos\Theta_{jk})\cos 2\Phi_{jk}}{r_{jk}^3} \right\rangle_{motion} \tag{96}$$

$$\Pi_2^n(\cos\beta')\cos n\alpha' = \sqrt{2/5}[A_{20n}\Pi_2^0(\cos\beta) + 2A_{22n}\Pi_2^2(\cos\beta)\cos 2\alpha] \tag{97}$$

and where various kinds of intra- and inter-molecular motions may be postulated, as described by Andrew,[116] Olf and Peterlin,[117] and Folkes and Ward.[118]

For the simplest case in the intra-molecular motions,[118] where molecular chains rotate freely around their chain axes, eqns (96) and (97) may be reduced as

$$A = \frac{\Pi_2^0(\cos\Theta_{jk})}{r_{jk}^3} \text{ and } B = 0 \tag{98}$$

Then, the second moment of NMR absorption line is given for the system of orthogonal–biaxial orientation as

$$\langle \Delta H^2 \rangle = \frac{G}{N} \frac{16}{25} \sum_{l=0}^{4} \sum_{m=-l}^{l} R_{lm} \Pi_l^m (\cos \beta) \cos (m\alpha) \tag{99}$$

where

$$R_{lm} = R_{l\bar{m}} = a_{lm} A_{lm0} \sum_{j>k} A^2 \tag{100}$$

and $a_{00} = \sqrt{2}/2$, $a_{20} = a_{40} = 2/7$, and $a_{22} = a_{42} = a_{44} = 4/7$. Finally, the averaged degree of orientation, A_{lm0} can be determined from eqns (98) through (100) by using the value of A given by eqn (98). Further, one can calculate the second moments of NMR absorption line for any other cases by postulating the models of the molecular motions, such as the harmonic, statistical, and Gaussian oscillations, and the flip-flop motion.[116] All of the molecular motions reduce the second moment, as observed as the line narrowing phenomena.

The fourth moment of NMR absorption line, $\langle \Delta H^4 \rangle$, has been also formulated with respect to the molecular orientation by several authors,[80,83,118] mostly according to the theory by Van Vleck,[80] to give higher orders of the orientation factors up to the eighth order. The discussions are, however, still problematic and somewhat too complicated to describe here comprehensively. The reader who is interested in the detail may refer to the above references, especially the paper by McBrierty and McDonald.[119]

5. GRAPHICAL REPRESENTATION OF THE STATE OF ORIENTATION OF STRUCTURAL ELEMENTS

As has been pointed out before, the orientation assessment of non-crystalline structural elements is only available from the absorption and/or emission dichroisms in terms of the second and/or fourth moments of the orientation distribution function of the $\mathbf{r}_j$ vectors, $q_j(\theta_j, \phi_j)$. This is quite a serious problem in assessing the non-crystalline orientation, especially when the orientation distribution becomes sharp, as discussed above, and necessitates more detailed discussion about the state of orientation in terms of these limited number of orientation factors up to the fourth orders. For simplicity, the problem will be discussed only for the uniaxial orientation of the jth vectors with respect of the x_3 axis.

Let us substitute the following relations given by

$$f(\zeta_j) = (\zeta_j)^r [q_j(\zeta_j)]^{1/2} \tag{101}$$

and

$$g(\zeta_j) = (\zeta_j)^s [q_j(\zeta_j)]^{1/2} \tag{102}$$

into the following Schwarz inequality:

$$\left[\int_{-1}^{1} f(\zeta_j) q_j(\zeta_j) \, d\zeta_j \right]^2 \leq \int_{-1}^{1} [f(\zeta_j)]^2 \, d\zeta_j \int_{-1}^{1} [g(\zeta_j)]^2 \, d\zeta_j \tag{103}$$

Then the relationship between the moments of the orientation distribution will be given, in general, by

$$(\langle \cos^{r+s} \theta_j \rangle)^2 \leq \langle \cos^{2r} \theta_j \rangle \langle \cos^{2s} \theta_j \rangle \tag{104}$$

For either $r = 2l$ and $s = 0$ or $r = 2l+1$ and $s = 1$, eqn (104) results in

$$(\langle \cos^{2l} \theta_j \rangle)^2 \leq \langle \cos^{4l} \theta_j \rangle \tag{105}$$

or

$$(\langle \cos^{2l+2} \theta_j \rangle)^2 / \langle \cos^2 \theta_j \rangle \leq \langle \cos^{4l+2} \theta_j \rangle \tag{106}$$

In addition, $|\cos \theta_j| \leq 1$, which means $\cos^{2l} \theta_j \geq \cos^{2l+2} \theta_j$, then

$$\langle \cos^{2l+2} \theta_j \rangle \leq \langle \cos^{2l} \theta_j \rangle \tag{107}$$

Consequently, the relationship between the second and fourth moments, i.e. $\langle \cos^2 \theta_j \rangle$ and $\langle \cos^4 \theta_j \rangle$, can be represented as

$$(\langle \cos^2 \theta_j \rangle)^2 \leq \langle \cos^4 \theta_j \rangle \leq \langle \cos^2 \theta_j \rangle \tag{108}$$

from which the relationship between f_{20}^j and f_{40}^j can be also deduced from eqns (29) and (31) as follows:

$$(5f_{20}^j + 7)/12 \geq f_{40}^j \geq 35[(f_{20}^j - 1/7)^2/18] - (3/7) \tag{109}$$

Therefore, any type of orientation distribution function involving uniaxial orientation must be plotted, as illustrated in Fig. 12, within the area defined by eqn (109) in the orthogonal coordinates of f_{20}^j and f_{40}^j. In turn, any type of orientation distribution may be assessed within the fourth order approximation, in principle, from comparison of the plots of f_{40}^j vs f_{20}^j with plots of certain characteristic orientation distribution functions.[95]

First of all, let us consider some extreme models of the uniaxial orientation, as illustrated in Fig. 13, where all of the vectors r_j orient uniaxially at

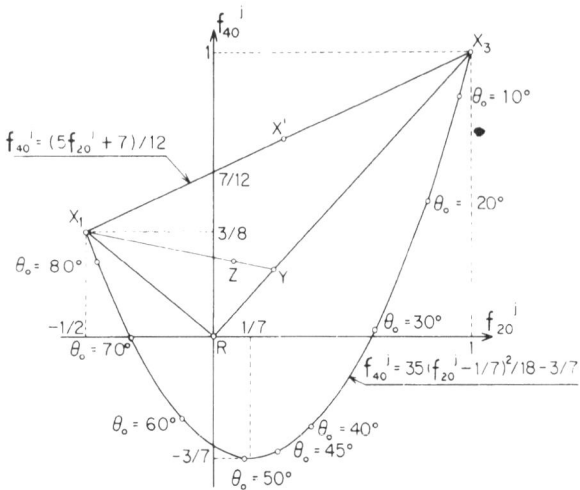

FIG. 12.   Orthogonal coordinates of f^i_{40} vs f^i_{20} representing uniaxial orientation of the $\mathbf{r}_j$ vectors. Any type of uniaxial orientations of the $\mathbf{r}_j$ vectors with respect to the x_3 axis must be plotted within the area. (From Nomura *et al.*)[95]

a given polar angle θ_0 with respect to the x_3 axis. For model I for which $\theta_0 = 0$, the plot degenerates to the point X_3 in Fig. 12. For model III with $\theta_0 = \pi/2$, the plot degenerates to the point X_1, while for model II for which $0 \leq \theta_0 \leq \pi/2$, the plot is given by the locus of the equation $f^i_{40} = 35[(f^i_{20} - 1/7)^2/18] - (3/7)$. Further, an extreme orientation distribution composed of two components, which are represented by the model I and the model III with relative amounts given by the ratio of straight-line segments $\overline{X_3 X'}$ and $\overline{X' X_1}$ in Fig. 12, is plotted at the point X'. The plot of another extreme orientation distribution composed of two components,

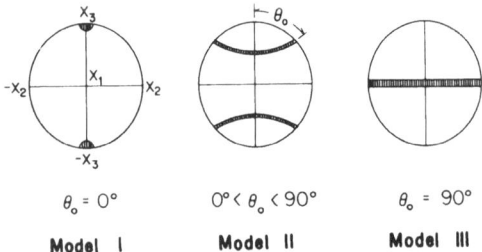

FIG. 13.   Pole figures illustrating three extremes concentrating the orientation distribution of the $\mathbf{r}_j$ vectors at the poles, a given polar angle θ_0, and equatorial zones all with respect to the x_3 axis; models I, II, and III, respectively. (From Nomura *et al.*)[95]

the model I and a random orientation (point R) in relative amounts $\overline{RY}$ and $\overline{YX_3}$, is given by the point Y, while an orientation distribution composed of two components, the model III and the orientation distribution represented by the point Y with relative amounts $\overline{YZ}$ and $\overline{ZX_1}$, corresponds to the point Z. Therefore, any plot of uniaxial orientation can be represented, in principle, by combinations of the models I, II, and III and random orientation with appropriate fractions of each component.

Similar considerations may be applied to the simple, but more realistic, distribution functions given by

$$q_j(\cos \theta_j) = N_a N_j(\theta_j) \tag{110}$$

where $N_j(\theta_j)$ is the distribution function of uniaxial orientation further defined for any of the following orientation models:

Model IV: $N_j(\theta_j) = (\cos^2 \theta_j)^a$ $\qquad\qquad$ (111)

Model V: $N_j(\theta_j) = (\sin^2 \theta_j)^a$ $\qquad\qquad$ (112)

Model VI: $N_j(\theta_j) = (2 \cos^2 \theta_0 \cos^2 \theta_j - \cos^4 \theta_j)^a$ $\qquad$ (113)

$$(0 < \theta_0 \leq \pi/4)$$

Model VII: $N_j(\theta_j) = (2 \sin^2 \theta_0 \sin^2 \theta_j - \sin^4 \theta_j)^a$ $\qquad$ (114)

$$(\pi/4 \leq \theta_0 < \pi/2)$$

Model VIII: $N_j(\theta_j) = [(\cos^2 \theta_j - \cos^2 \theta_0)^2]^a$ $\qquad$ (115)

$$(0 < \theta_0 < \pi/2)$$

where N_a is a normalisation constant given by

$$1/N_a = 2\pi \int_0^\pi N_j(\theta_j) \sin \theta_j \, d\theta_j \tag{116}$$

The function given by eqn (111) or (112) has a single maximum when θ_j is 0 or $\pi/2$, respectively. The functions given by eqns (113) and (114) also give a single maximum at $\theta_j = \theta_0$, while eqn (115) gives maxima at $\theta_j = 0$ and $\pi/2$ and a minimum at $\theta_j = \theta_0$. These features of the respective models are illustrated in Fig. 14 taking the value of a as unity or two. The plot of f_{40}^i vs f_{20}^i for each of the orientation models in Fig. 14 is depicted in Fig. 15 by an open circle or dot with the same notation as that in Fig. 14.

When $a \to 0$, every function defined by eqns (111) through (116) represents the random orientation corresponding to the point R in Fig. 15. On the other hand, when $a \to \infty$, the function defined by eqns (111) and (112)

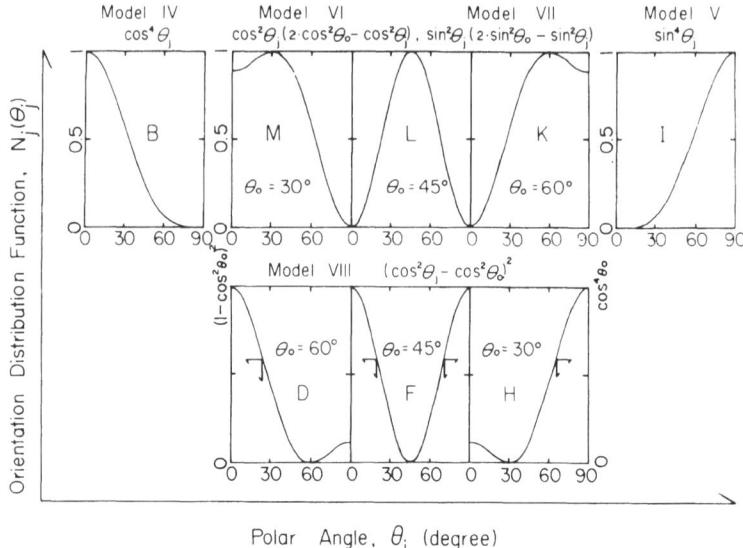

FIG. 14.   Some examples of distribution functions for uniaxial orientation corresponding to model IV–VIII given by eqns (111)–(115), where a is chosen as unity or 2. (From Nomura *et al.*)[95]

represents the models I and III, which correspond to the points X_3 and X_1 in Fig. 15, respectively. Equation (115) with $a \to \infty$ represents the model I or III, depending on the choice of θ_0, and eqns (113) and (114) with $a \to \infty$ represent the model II.

If the value of a varies from 0 to ∞, the plot of f_{40}^i vs f_{20}^i for the model IV, eqn (111), gives curve *RABX$_3$* in Fig. 15, where $f_{40}^i = f_{20}^i \times (5f_{20}^i - 2)/(5 - 2f_{20}^i)$, while the plot for the model V, eqn (112), gives curve *RJIX$_1$*, where $f_{40}^i = 3f_{20}^i(5f_{20}^i + 1)/(8f_{20}^i + 10)$. The corresponding plots for the model VI, eqn (113), with $\theta_0 = 30, 40, 45°$ are represented by curves *RMW*, *RV*, and *RLU*, respectively. Plots for the model VII, eqn (114), with $\theta_0 = 45, 50, 60°$ are curves *RLU*, *RT*, and *RKS*, respectively. Plots for the model VIII, eqn (115), with $\theta_0 = 30, 40, 45, 50, 60°$ are curves *RHX$_1$*, *RGX$_1$*, *RFX$_1$*, *REX$_3$*, and *RDX$_3$*, respectively.

Plots of f_{40}^i vs f_{20}^i for models such as those given by eqn (115) or the combination of model I and model III, which have double maxima at $\theta_j = 0$ and $\pi/2$, fall in the region above the curves X_1JR and *RABX$_3$* in Fig. 15, while the plots for models such as those given by eqns (113) and (114) and model II, which have a single maximum at $\theta_j = \theta_0 (0 < \theta_0 < \pi/2)$ fall

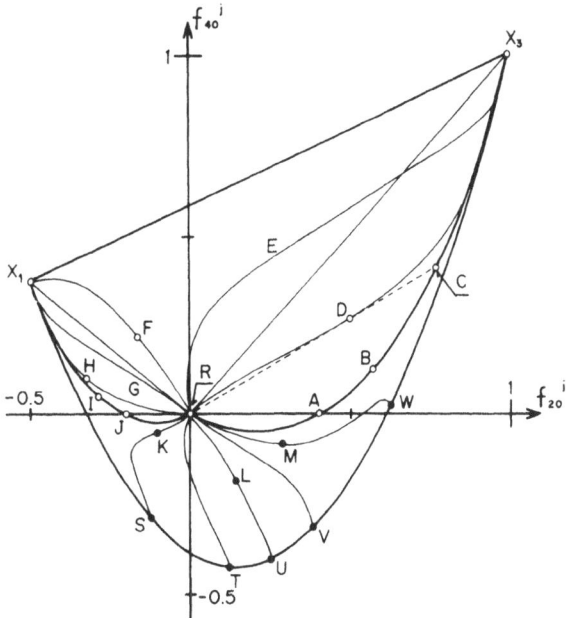

FIG. 15. Plots of f_{40}^j vs f_{20}^j for models IV–VIII for $0 \le a \le \infty$ and θ_0 fixed as
indicated. (From Nomura *et al.*)[95]

in the region below these curves. Point D in Fig. 15, the intersection of the
curve RDX_3 and the straight line RDC, corresponds to a distribution
function given by eqn (115) or to an extreme distribution involving a
combination of random orientation and the orientation corresponding to
the point C with relative amounts of $\overline{CD}$ and $\overline{DR}$, respectively. In other
words, the point D may arise from two indistinguishable types of uniaxial
orientation. However, the distribution function for uniaxial orientation of
non-crystalline chain segments usually has a single maximum at $\theta_j = 0$;
hence this ambiguity may not be of practical concern.

Let us further discuss the assessment of uniaxial orientation of non-
crystalline structural elements in comparison with the plots of f_{20}^j and f_{40}^j of
some specific orientation distribution functions. The following four specific
distribution functions will be considered, first:

Model IX: $\quad q_j(\cos \theta_j) = (1/4\pi) \dfrac{\lambda^3}{[\lambda^3 - (\lambda^3 - 1) \cos^2 \theta_j]^{3/2}}$ (117)

Model X: $\quad q_j(\cos \theta_j) = (N_b/2\pi)[\exp(\cos^2 \theta_j)]^b$ (118)

Model XI: $q_j(\cos \theta_j) = (N_\lambda/2\pi) \left\{ \dfrac{\lambda^3}{[\lambda^3 - (\lambda^3 - 1)\cos^2 \theta_j]^{3/2}} - \dfrac{1}{\lambda^{9/2}} \right\}$

$$(119)$$

Model XII: $q_j(\cos \theta_j) = (N_c/2\pi)\{[\exp(\cos^2 \theta_j)]^c - 1\}$ (120)

Here, N_b, N_λ, and N_c are normalisation constants and λ is the extension ratio of affine matrix. Equation (117) is the distribution function proposed by Kratky[16] for his floating rod model in an affine matrix and was already given by eqn (40). Equations (119) and (120) are deduced by subtracting the random component from eqns (117) and (118), respectively.

Variation of b and c or λ from 0 to ∞ or from unity to ∞ gives plots of f^i_{40} vs f^i_{20} represented by curves IX, X, XI, and XII in Fig. 16 for the respective models. Curve IV for the model IV given by eqn (111), is added for

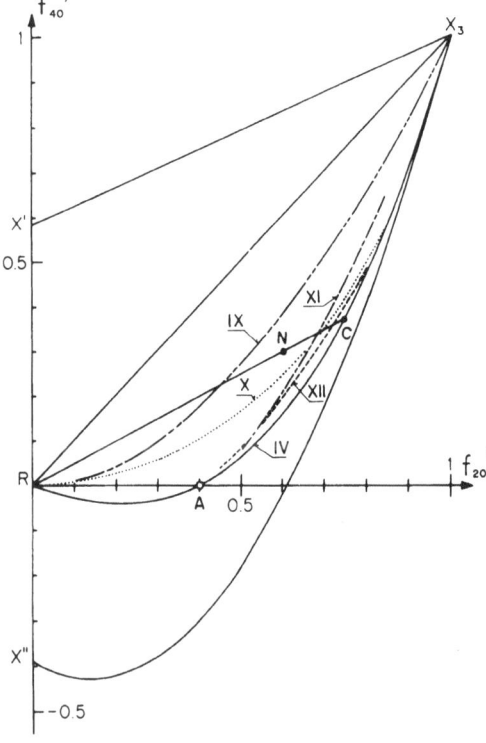

FIG. 16.   Plots of f^i_{40} vs f^i_{20} for four uniaxial orientation distributions, models IX–XII, given by eqns (117)–(120), respectively, with b and c between zero and infinity, and λ between unity and infinity. (From Nomura *et al.*)[95]

comparison. The distribution function corresponding to the curve IX is the sharpest of the four functions, as shown by the fact that it falls closest to the line RX_3 being composed of random orientation and the model I. With increasing f^i_{20}, curves X and XII converge to the point X_3 and approximate curve IV more closely than curves IX and XI. With decreasing f^i_{20}, on the other hand, curves IX and X converge to the point R (random orientation), while curves XI and XII approximate point A.

From the model of Kuhn and Grün of the so-called freely jointed equivalent chain[44] it is worthwhile examining the orientation behaviour of crosslinked polymer networks in the rubbery state in terms of the plots of f^i_{40} vs f^i_{20} according to the procedure by Roe and Krigbaum.[120] For a rubber network chain consisting of N freely jointed statistical segments, the $2n$th moments of the end-to-end vector distance r_0^{2n} in the unstrained state may be given by

$$r_0^2 = \gamma N b^2$$
$$r_0^4 = \gamma^2 (5/3) N^2 b^4 \qquad (121)$$
$$r_0^6 = \gamma^3 (35/9) N^3 b^6$$

where b is the length of the statistical segment, and γ is a parameter characterising the ratio of the end-to-end vector distance of the real rubber chain to that of the freely jointed chain, both in the unstrained state.

On deformation, the end-to-end vectors undergo an affine transformation under constant volume, that is, the orientation distribution of the end-to-end vectors with respect to the stretching direction is given, similarly to eqn (117), by

$$g_\lambda(\alpha) = (1/2)(r/r_0)^3 \sin \alpha = (1/2)\frac{\lambda^3 \sin \alpha}{[\lambda^3 - (\lambda^3 - 1)\cos^2\alpha]^{3/2}} \qquad (122)$$

where α is the angle between the end-to-end vector and the stretching direction and r is the end-to-end vector distance under the deformation as given by

$$r^2 = r_0^2 \frac{\lambda^2}{\lambda^3 - (\lambda^3 - 1)\cos^2\alpha} \qquad (123)$$

On the other hand, the orientation distribution of the statistical segments with respect to the end-to-end vector can be given by

$$f(\delta) = (1/2)(\beta/\sinh \beta) \exp(\beta\delta) \qquad (124)$$

where δ is the cosine of the angle between the statistical segment and the end-to-end vector, and β is the inverse Langevin function of (r/Nb), i.e.

$$\beta = L^{-1}(r/Nb)$$

or

$$(r/Nb) = \coth \beta - 1/\beta = L(\beta) \qquad (125)$$

Thus, the orientation factor of the statistical segments with respect to the stretching direction can be given in general form as

$$f_{l00} = \int_{-1}^{1} \left[\int_{-1}^{1} f(\delta) P_l(\delta) \, d\delta \right] g_\lambda(\alpha) P_l(\cos \alpha) \, d\alpha \qquad (126)$$

The second- and fourth-order orientation factors are given by

$$f_{200} = (1/5)(\gamma/N)(\lambda^2 - \lambda^{-1}) + (36/875)(5/3)(\gamma/N)^2(\lambda^4 + \lambda/3 - 4/3\lambda^2)$$
$$+ (108/6125)(35/9)(\gamma/N)^3(\lambda^6 + 3\lambda^2/5 - 8/5\lambda^3) \qquad (127)$$

$$f_{400} = (3/175)(5/3)(\gamma/N)^2(\lambda^4 - 2\lambda + 1/\lambda^2)$$
$$+ (216/13475)(35/9)(\gamma/N)^3(\lambda^6 - 4\lambda^3/5 - 7/5 + 6/5\lambda^3) \qquad (128)$$

In Fig. 17 are shown the changes of the second- and fourth-order orientation factors with %-elongation, i.e. $(\lambda - 1) \times 100\%$, for various

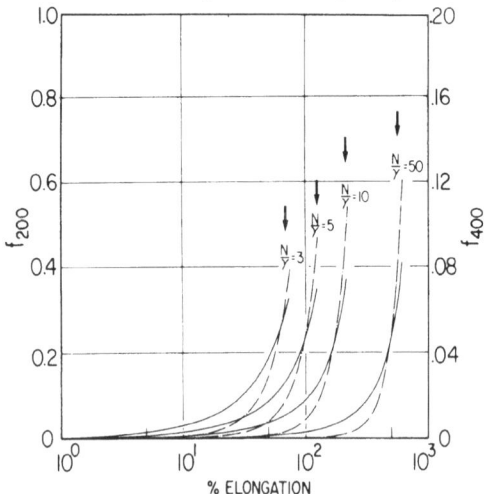

FIG. 17. Changes of orientation factors, f_{200} and f_{400}, for the statistical segments in the model of Kuhn and Grün with %-elongation and N/γ fixed as indicated. ——, f_{200}; – · – · –, f_{400}.

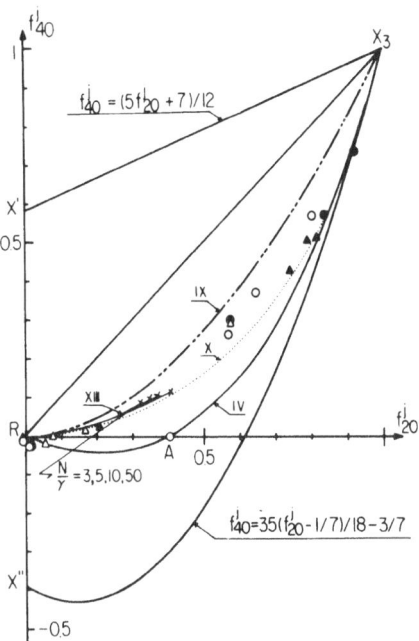

FIG. 18. Comparison of plots of f_{400} vs f_{200} for the model of Kuhn and Grün with those of f_{40}^j vs f_{20}^j for the models IV, IX, and X, where λ varies between unity and $(N/\gamma)^{1/2}$ for the model of Kuhn and Grün. Some experimental results for the uniaxial orientation behaviour of poly(ethylene terephthalate) are added. Circle and triangle symbols represent uniaxial orientation behaviour of crystalline and non-crystalline molecular axes, respectively, for the stretched specimens (open symbols) and the specimens annealed with respective fixed lengths (dark symbols). (From Yoshihara *et al.*)[121]

values of (N/γ) ranging from 3 to 50, in which each arrow along the respective curves indicates an upper limit of the elongation at which the network chain, whose end-to-end vector orients parallel to the stretching direction, is fully extended; i.e. $\lambda = (N/\gamma)^{1/2}$. As can be seen in the figure, the larger the value of (N/γ), the more gradually the segments orient with %-elongation.

Figure 18 further shows the orientation behaviour of the model of Kuhn and Grün, which is designated as model XIII, in terms of the f_{20}^j vs f_{40}^j plots in comparison with those of the other models, such as models IV, IX, and X. It can be seen that the behaviour is rather insensitive to the value of (N/γ), giving nearly a common curve along which the cross-marks correspond to the upper limits of the orientation with $(N/\gamma) = 3, 5, 10, 50,$

respectively. It is also noted that the common curve locates more apart from the straight line of RX_3 and closer to the curve X for model X, both than the curve IX for model IX. It is suggested that the orientation distribution of the model of Kuhn and Grün is considerably broader than that of model IX and changes very gradually with an increase of λ, as does model X. The circular and triangular symbols plotted together in the figure show the uniaxial orientation behaviour of crystalline and non-crystalline molecular axes of quenched poly(ethylene terephthalate), respectively, during stretching up to a stretch-ratio of 4 (open symbols) and after annealing the stretched samples at 180°C with respective fixed lengths (dark symbols).[121] It can be noted that the uniaxial orientation behaviour of this particular poly(ethylene terephthalate) sample is rather mild in the initial stage of stretching, becomes considerably more drastic in the intermediate stretching range, and is mild again in the final stage of stretching, probably resulting from orientation and disintegration mechanisms of some crystalline texture, if any.

6. DEFORMATION MECHANISM OF SPHERULITIC CRYSTALLINE TEXTURE

As one of the applications of the theory of Krigbaum and Roe interrelating two types of orientation distribution functions through eqn (11), let us show an investigation of the deformation mechanism of spherulitic crystalline texture of polyethylene based on a model illustrated in Fig. 6. The model postulates two types of preferential deformation mechanisms of crystal lamellae in association with lamellar orientation during uniaxial deformation of the spherulitic texture, i.e. untwisting of a lamella around the lamellar axis (rotation of the crystal a and c axes around the crystal b axis) and tilting of the crystal c axis toward the [100] direction (rotation of crystal b and c axes around the crystal a axis) at the equatorial and polar zones of the deformed spherulites, respectively, both due to straining of tie-chain molecules between adjacent lamellae. The tilting mechanism is, however, postulated simply as given by a fixed angle of κ, not as being of any further function of lamellar orientation angle θ or of spherulite extension ratio λ. Therefore, the problem is simply calculating $q_j(\zeta_j)$ for a given $\mathbf{r}_j$ fixed within the lamella with given angles of Θ_j and Φ_j from $w(\xi, \phi, \eta)$ of the lamellar orientation by utilising the relation given by eqn (20), in principle.

The orientation distribution function of the crystal lamellae within the affine matrix, in combination with the untwisting mechanism, may be formulated as:[122]

$$w_{\text{lamella}}(\xi,0,\eta) = N(\xi)[1+\sigma \cdot g(\xi)\cos 2\eta] \qquad (129)$$

where $N(\xi)$ may be the same distribution function as that given by eqn (117), σ is the parameter of the ease of lamellar untwisting, and g is a function characterising the untwisting mechanism, as proposed by Stein *et al.*,[109] such that

$$g = 2\langle\cos^2\eta\rangle - 1$$
$$= (\lambda - 1)\sin^2\theta = (\lambda - 1)(1 - \xi^2) \qquad (130)$$

In other words, the degree of untwisting is formulated not only to be proportional to the spherulite deformation $(\lambda - 1)$, but also to be the most significant within the lamellae at the equatorial zone of $\theta = 90°$.

The coefficients of the expanded series of the distribution function of the lamellar orientation may be given by

$$2\pi A_{l00} = \int_0^{2\pi} \int_{-1}^{1} Z_{l00}(\xi)N(\xi)\,d\xi\,d\eta \qquad (131)$$

$$2\pi A_{l02} = \int_0^{2\pi} \int_{-1}^{1} Z_{l02}(\xi)\sigma \cdot g(\xi)N(\xi)\cos^{2\eta}d\xi\,d\eta \qquad (132)$$

and the other $A_{l0n} = 0$.

The distribution function, $q_j(\zeta_j)$ of the $|\mathbf{r}_j$vector may be given by eqn (19) as follows:

$$q_j(\zeta_j) = \sum_{l=0}^{\infty} \alpha_{l0}^j \Pi_l(\zeta_j) \qquad (133)$$

TABLE 2

GEOMETRIC RELATIONSHIPS OF THE THREE PRINCIPAL CRYSTALLOGRAPHIC AXES OF POLYETHYLENE CRYSTAL (ORTHORHOMBIC)

	$\cos \Theta_j$	$\cos 2\Phi_j$
a axis	0	-1
b axis	$\cos\kappa$	1
c axis	$-\sin\kappa$	
$U_{(110)}$	$\cos\kappa\cos 33\cdot69°$	$\dfrac{\sin^2\kappa - \tan^2 33\cdot69°}{\sin^2\kappa + \tan^2 33\cdot69°}$

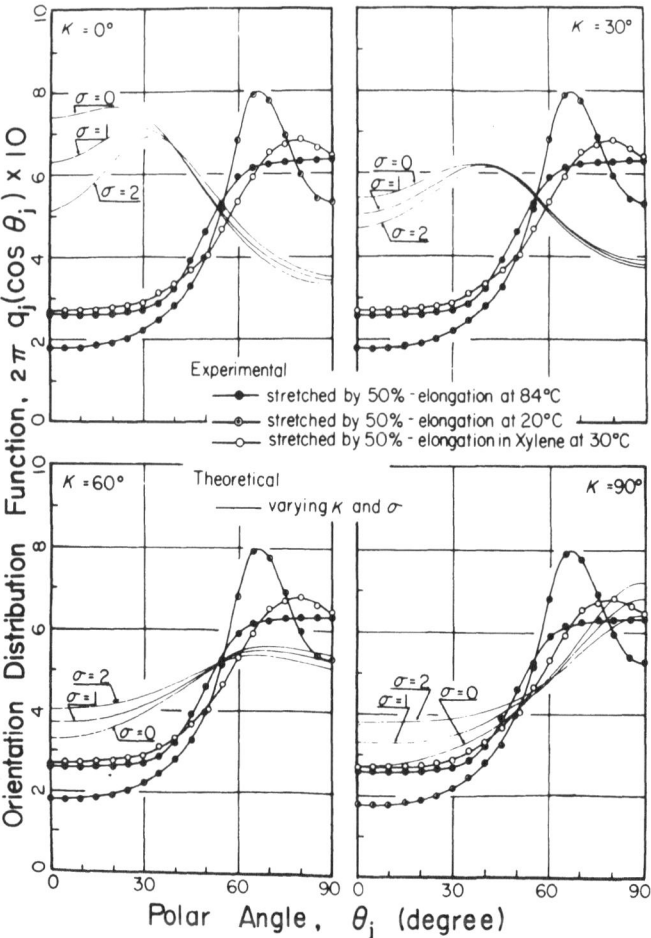

FIG. 19. Comparison of the orientation distribution function of the (110) crystal planes observed from X-ray diffraction for a low-density polyethylene stretched uniaxially to 50%-elongation under various environmental conditions, with those calculated from eqns (131) and (132) by varying the untwisting parameter σ and the tilting angle κ. (From Nomura *et al.*)[122]

where α_{l0}^{j} is the coefficient of the expanded series of $q_j(\zeta_j)$ and can be further given from eqn (20) as follows:

$$\alpha_{l0}^{j} = 2 \left(\frac{2}{2l+1} \right)^{1/2} [A_{l00}\Pi_l(\cos\Theta_j) + 2A_{l02}\Pi_l^2(\cos\Theta_j)\cos 2\Phi_j] \quad (134)$$

The geometric relationships of the three principal crystallographic axes of a polyethylene crystal, the a, b, and c axes, and the reciprocal lattice vector of the (110) crystal plane, $U_{(110)}$, within the lamella, are listed in Table 2 in terms of the angles, Θ_j and Φ_j, and the tilting angle κ toward the [100] direction of the original lamella.

Figure 19 shows comparison of the azimuthal distributions of X-ray diffraction intensity from the (110) crystal plane, the most intensive reflection, for a low density polyethylene stretched to 50%-elongation under various environmental conditions, with the distributions calculated from eqns (133) and (134) by varying the untwisting parameter σ and the tilting angle κ.[122] Although there exist considerable discrepancies between the calculated and observed results, it can be understood that the tilting angle κ should be taken as being larger than, at least, 60°, and that the effect of the untwisting is rather minor in this model. The discrepancies could be reduced if one takes the tilting angle κ as a function of the lamellar orientation angle θ and the spherulite deformation λ. More advanced studies along this line have been carried out by the present authors elsewhere.[107, 123–126]

REFERENCES

1. NEUBERT, H., *Kolloidchem. Beihefte*, 1925, **20**, 244.
2. FREY, A., *Naturwiss.*, 1925, **13**, 403.
3. FREY, A., *Kolloidchem. Beihefte*, 1925, **20**, 209.
4. AMBRONN, H., *Ber. deut. botan. Ges.*, 1888, **6**, 85, 225.
5. WIENER, O., *Sächs. Ges. Wiss. Leipzig. Math.–Phys. Kl. Abh.*, 1912, **32**, 509.
6. WIENER, O., *Kolloidchem. Beihefte*, 1927, **23**, 189.
7. MÖHRING, A., *Kolloidchem. Beihefte*, 1927, **23**, 162.
8. PRESTON, J. M., *J. Soc. Dyers Col.*, 1931, **47**, 312.
9. PRESTON, J. M. and TSIEN, P. C., *J. Soc. Dyers Col.*, 1946, **62**, 242.
10. PRESTON, J. M. and TSIEN, P. C., *J. Soc. Dyers Col.*, 1946, **62**, 368.
11. PRESTON, J. M. and SU, Y. F., *J. Soc. Dyers Col.*, 1950, **66**, 375.
12. PRESTON, J. M. and TSIEN, P. C., *J. Soc. Dyers Col.*, 1950, **66**, 361.
13. MOREY, D. R., *Textile Res.*, 1933, **3**, 325.
14. MOREY, D. R., *Textile Res.*, 1934, **4**, 491.
15. MOREY, D. R., *Textile Res.*, 1935, **5**, 105, 483, 538.
16. KRATKY, O., *Kolloid Z.*, 1933, **64**, 213.
17. KRATKY, O., *Kolloid Z.*, 1934, **68**, 347.
18. KRATKY, O., *Kolloid Z.*, 1935, **70**, 14.
19. BREUER, F., KRATKY, O. and SAITO, C., *Kolloid Z.*, 1937, **80**, 139.
20. KRATKY, O., *Kolloid Z.*, 1938, **84**, 149.
21. KRATKY, O. and PLATZEK, P., *Kolloid Z.*, 1938, **84**, 268.

22. HERMANNS, P. H., KRATKY, O. and PLATZEK, P., *Kolloid Z.*, 1939, **86**, 245.
23. KRATKY, O. and PLATZEK, P., *Kolloid Z.*, 1939, **88**, 78.
24. KRATKY, O., *Angew. Chem.*, 1940, **53**, 153.
25. BAULE, B., KRATKY, O. and TREER, R., *Z. Phys. Chem.*, 1941, **B50**, 255.
26. BAULE, B. and KRATKY, O., *Z. Phys. Chem.*, 1942, **B52**, 142.
27. HERMANNS, P. H., *Kolloid Z.*, 1937, **81**, 143.
28. HERMANNS, P. H. and LEEUW, A. J., *Kolloid Z.*, 1937, **81**, 300.
29. HERMANNS, P. H. and LEEUW, A. J., *Kolloid Z.*, 1938, **82**, 58.
30. HERMANNS, P. H., *Kolloid Z.*, 1938, **83**, 71.
31. HERMANNS, P. H., *Kolloid Z.*, 1939, **86**, 107.
32. HERMANNS, P. H., *Rec. Trav. Chim.*, 1939, **58**, 63.
33. HERMANNS, P. H. and PLATZEK, P., *Kolloid Z.*, 1939, **87**, 296.
34. HERMANNS, P. H., and PLATZEK, P., *Kolloid Z.*, 1939, **88**, 68.
35. HERMANNS, P. H. and BOOYS, J., *Kolloid Z.*, 1939, **88**, 73.
36. HERMANNS, P. H., *Kolloid Z.*, 1939, **88**, 172.
37. HERMANNS, P. H., *Proc. Kon. Nedel. Akad. Wetensch.*, 1939, **42**, 798.
38. HERMANNS, P. H., *Kolloid Z.*, 1939, **89**, 344.
39. HERMANNS, P. H. and PLATZEK, P., *Kolloid Z.*, 1939, **89**, 349.
40. HERMANNS, P. H. and PLATZEK, P., *Z. Phys. Chem.*, 1939, **A185**, 260.
41. BOOYS, J., BREDEE, H. L. and HERMANNS, P. H., *Rec. Trav. Chim.*, 1940. **59**, 73.
42. HERMANNS, P. H., *Naturwiss.*, 1940, **28**, 223.
43. MÜLLER, F. H., *Kolloid Z.*, 1941, **95**, 138, 306.
44. KUHN, W. and GRÜN, F., *Kolloid Z.*, 1942, **101**, 248.
45. TRELOAR, L. R. G., *Trans. Faraday Soc.*, 1947, **43**, 277, 284; 1954, **50**, 881, *The Physics of Rubber Elasticity, 2nd Ed.*, 1958, Oxford Univ. Press, London.
46. RAUMANN, G. and SAUNDERS, D. W., *Proc. Phys. Soc.*, 1961, **77**, 1028.
47. RAUMANN, G., *Proc. Phys. Soc.*, 1962, **79**, 1221.
48. RAUMANN, G., *Brit. J. Appl. Phys.*, 1963, **14**, 795.
49. WARD, I. M., *Proc. Phys. Soc.*, 1962, **80**, 1176.
50. HADLEY, D. W., WARD, I. M. and WARD, J., *Proc. Roy. Soc. (London)*, 1966, **A285**, 275.
51. PINNOCK, P. R., WARD, I. M. and WOLFE, J. M., *Proc. Roy. Soc. (London)*, 1966, **A291**, 267.
52. WARD, I. M., *Appl. Mat. Res.*, 1966, **5**, 224, 228.
53. LADIZEWSKY, N. H. and WARD, I. M., *J. Macromol. Sci.–Phys.*, 1971, **B5**, 661, 745.
54. GÜPTA, V. B., and WARD, I. M., *J. Macromol. Sci.–Phys.*, 1967, **B1**, 373.
55. STACHURSKI, Z. H. and WARD, I. M., *J. Polym. Sci.*, A–2, 1968, **6**, 1083, 1817; *J. Macromol. Sci. Phys.*, 1969, **B3**, 427, 445.
56. GUPTA, V. B. and WARD, I. M., *J. Macromol. Sci.–Phys.*, 1968, **B2**, 89.
57. GUPTA, V. B., KELLER, A. and WARD, I. M., *J. Macromol. Sci.–Phys.*, 1968, **B2**, 139.
58. NOMURA, S., KAWABATA, S., KAWAI, H., YAMAGUCHI, Y., FUKUSHIMA, A. and TAKAHARA, H., *J. Polym. Sci.*, A–2, 1969, **7**, 325.
59. TAKAHARA, H., NOMURA, S., KAWAI, H., YAMAGUCHI, Y., OKAZAKI, K. and FUKUSHIMA, A., *J. Polym. Sci.*, A–2, 1968, **6**, 197.

60. MAEDA, M., HIBI, S., ITO, F., NOMURA, S., KAWAGUCHI, T. and KAWAI, H., *J. Polym. Sci.*, *A–2*, 1970, **8**, 1303.
61. NOMURA, S., KAWAI, H., KIMURA, I. and KAGIYAMA, M., *J. Polym. Sci.*, *A–2*, 1967, **5**, 479.
62. FRASER, R. D. B., *J. Chem. Phys.*, 1953, **21**, 1511; 1958, **28**, 1113; 1958, **29**, 1428.
63. BEER, M., *Proc. Roy. Soc. (London)*, 1956, **A236**, 136.
64. STEIN, R. S., *J. Appl. Phys.*, 1961, **32**, 1280.
65. KOENIG, J. L., CORNELL, S. W. and WITENHAFER, D. E., *J. Polym. Sci.*, *A–2*, 1967, **5**, 301.
66. OKAJIMA, S., KOBAYASHI, Y. and KAWATA, Y., *Kogyo Kagaku Zasshi*, 1956, **59**, 1213.
67. SHINDO, Y. and STEIN, R. S., *J. Polym. Sci.*, 1967, **B5**, 737.
68. SHINDO, Y., READ, B. E. and STEIN, R. S., *Macromol. Chem.*, 1968, **118**, 272.
69. NISHIJIMA, Y., ONOGI, Y. and ASAI, T., *Rep. Prog. Polym. Phys.*, *Japan*, 1965, **8**, 131.
70. NISHIJIMA, Y., ONOGI, Y. and ASAI, T., *J. Polym. Sci.*, 1966, **C15**, 237.
71. ONOGI, Y. and NISHIJIMA, Y., *Rep. Prog. Polym. Phys.*, *Japan*, 1976, **19**, 411, 415, 419.
72. DESPER, C. R. and KIMURA, I., *J. Appl. Phys.*, 1967, **38**, 4225.
73. KIMURA, I., KAGIYAMA, M., NOMURA, S. and KAWAI, H., *J. Polym. Sci.*, *A–2*, 1969, **7**, 709.
74. ROE, R. J., *J. Polym. Sci.*, *A–2*, 1970, **8**, 1187.
75. BOWER, D. I., *J. Polym. Sci.*, *Polym. Phys. Ed.*, 1972, **10**, 2135.
76. HIBI, S., MAEDA, M., KUBOTA, H. and MIURA, T., *Polymer*, 1977, **18**, 143.
77. HIBI, S., FUJITA, K., MAEDA, M., NODA, A., SUZUKI, M. and OZAKI, M., *Sen-i Gakkai-shi (J. Soc. Fibre Sci. & Tech., Japan)*, 1981, **37**, T215.
78. CORNELL, S. W. and KOENIG, J. L., *J. Appl. Phys.*, 1968, **39**, 4883.
79. SNYDER, R. G., *J. Mol. Spectr.*, 1971, **37**, 353.
80. VAN VLECK, J. H., *Phys. Rev.*, 1948, **74**, 1168.
81. YAMAGATA, K. and HIROTA, S., *J. Appl. Phys.*, *Japan*, 1961, **30**, 261.
82. MCBRIERTY, V. J. and WARD, I. M., *Brit. J. Appl. Phys. (J. Phys. D.)*, 1968, **1**, 1529.
83. MCBRIERTY, V. J., MCDONALD, I. R. and WARD, I. M., *J. Phys. D: Appl. Phys.*, 1971, **4**, 88.
84. MCBRIERTY, V. J., *Polymer*, 1974, **15**, 503.
85. MCBRIERTY, V. J. and MCDONALD, I. R., *Polymer*, 1975, **16**, 125.
86. SAMUELS, R. J., in: *The Science and Technology of Polymer Films*, O. J. Sweeting, Ed., 1968, Wiley, New York, pp. 255–364; Samuels, R. J., *Structured Polymer Properties*, 1974, Wiley, New York.
87. WILKES, G. L., *Adv. Polym. Sci.*, 1971, **8**, 91.
88. WARD, I. M., *Structures and Properties of Oriented Polymers*, 1975, Applied Science, London.
89. ROE, R. J. and KRIGBAUM, W. R., *J. Chem. Phys.*, 1964, **40**, 2608.
90. ROE, R. J., *J. Appl. Phys.*, 1965, **36**, 2024.
91. NOMURA, S., KAWAI, H., KIMURA, I. and KAGIYAMA, M., *J. Polym. Sci.*, *A–2*, 1970, **8**, 383.

262 HIROMICHI KAWAI AND SHUNJI NOMURA

92. STEIN, R. S., *J. Polym. Sci.*, 1958, **31**, 327.
93. DESPER, C. R. and STEIN, R. S., *J. Appl. Phys.*, 1966, **37**, 3990.
94. DESPER, C. R., *J. Appl. Phys.*, 1968, **39**, 344.
95. NOMURA, S., NAKAMURA, N. and KAWAI, H., *J. Polym. Sci.*, *A-2*, 1971, **9**, 407.
96. STEIN, R. S., and READ, B. E., in: *International Symposium on Polymer Characterization* (*Appl. Polym. Symp.*, 8), K. A. Boni and F. A. Sliemers, Eds, 1969, Interscience, New York, p. 255.
97. HEIKENS, D., *Dichroism of Dyed Cellulose Fibers*, 1952, F. A. Schotanus and Jens, Utrecht.
98. ZBINDEN, R., *Infrared Spectroscopy of High Polymers*, 1964, Academic Press, New York, Ch. 5.
99. STEIN, R. S., *J. Polym. Sci.*, 1958, **31**, 327, 335.
100. ALEXANDER, L. E., *X-ray Diffraction Methods in Polymer Science*, 1969, Wiley, New York, Ch. 4.
101. STEIN, R. S. and NORRIS, F. H., *J. Polym. Sci.*, 1956, **21**, 381.
102. WILCHINSKI, Z. W., *J. Appl. Phys.*, 1959, **30**, 792.
103. SACK, R. A., *J. Polym. Sci.*, 1961, **54**, 543.
104. KRIGBAUM, W. R. and ROE, R. J., *J. Chem. Phys.*, 1964, **41**, 737.
105. KRIGBAUM, W. R., ADACHI, T. and DAWKINS, J. V., *J. Chem. Phys.*, 1968, **49**, 1532.
106. TAKAHARA, H., *PhD Thesis* Kyoto University, 1969; Takahara, H., Kawai, H., Fukushima, T. and Yamaguchi, Y., *Seni-i Gakkai-shi* (*J. Soc. Fibre Sci. & Tech., Japan*), 1971, **27**, 338.
107. FUJITA, K., SUEHIRO, S., NOMURA, S. and KAWAI, H., *Polymer J., Japan.*, 1982, **14**, 545.
108. SASAGURI, K., HOSHINO, S. and STEIN, R. S., *J. Appl. Phys.*, 1964, **35**, 47.
109. SASAGURI, K., YAMADA, R. and STEIN, R. S., *J. Appl. Phys.*, 1964, **35**, 3188.
110. FUJINO, K., KAWAI, H., ODA, T. and MAEDA, H., *Proc. 4th Intern. Congr. Rheology*, 1965, Part 3, 501.
111. ODA, T., NOMURA, S. and KAWAI, H., *J. Polym. Sci.*, 1965, **A3**, 1993.
112. ODA, T., SAKAGUCHI, N. and KAWAI, H., *J. Polym. Sci.*, 1966, **C15**, 223.
113. NOMURA, S., *PhD Thesis*, Kyoto University, 1970.
114. STEIN, R. S., *J. Polym. Sci.*, 1957, **24**, 383.
115. O'REILLY, D. E. and TSANG, T., *Phys. Rev.*, 1962, **128**, 2639.
116. ANDREW, E. R., *J. Chem. Phys.*, 1950, **18**, 607.
117. OLF, H. G. and PETERLIN, A., *J. Polym. Sci.*, *A-2*, 1970, **8**, 753.
118. FOLKES, M. J. and WARD, I. M., *J. Mater. Sci.*, 1971, **6**, 582.
119. McBRIERTY, V. J. and McDONALD, I. R., *J. Phys. D: Appl. Phys.*, 1973, **6**, 131.
120. ROE, R. J. and KRIGBAUM, W. R., *J. Appl. Phys.*, 1964, **35**, 2215.
121. YOSHIHARA, N., FUKUSHIMA, A., WATANABE, Y., NAKAI, A., NOMURA, S. and KAWAI, H., *J. Soc. Fibre Sci. & Tech., Japan* (*Sen-i Gakkai-shi*), 1981, **37**, T387.
122. NOMURA, S., ASANUMA, A., SUEHIRO, S. and KAWAI, H., *J. Polym. Sci.*, *A-2*, 1971, **9**, 1991.
123. NOMURA, S., MATSUO, M. and KAWAI, H., *J. Polym. Sci., Polym. Phys. Ed.*, 1972, **10**, 2489.

124. MATSUO, M., NOMURA, S. and KAWAI, H., *J. Polym. Sci., Polym. Phys. Ed.*, 1973, **11**, 2057.
125. NOMURA, S., MATSUO, M. and KAWAI, H., *J. Polym. Sci., Polym. Phys. Ed.*, 1974, **12**, 1371.
126. MATSUO, M., HATTORI, H., NOMURA, S. and KAWAI, H., *J. Polym. Sci., Polym. Phys. Ed.*, 1976, **14**, 223.

INDEX